T0167408

RAND

Limiting the Spread of Weapon-Usable Fissile Materials

Brian G. Chow,
Kenneth A. Solomon

Prepared for the
Under Secretary of Defense for Policy

**National Defense
Research Institute**

PREFACE

This report examines the problem of rapidly accumulating weapon-usable fissile materials and proposes an agenda to help the United States and other countries manage these materials. Weapon-usable fissile materials come from both dismantled nuclear weapons and the spent fuel from civilian nuclear power plants. This report should be of interest to nuclear nonproliferation planners and analysts in the United States, the former Soviet republics (FSRs), and other countries, and also to nuclear energy planners.

The study started in October 1991. By June 1992, we had briefed our interim recommendations to planners and analysts in various DoD offices and also in the National Security Council, Livermore National Laboratory, and the nuclear industry. We also solicited reactions from public interest groups, particularly on the ramifications of our key recommendation: that the United States purchase highly enriched uranium from the FSRs after it is diluted and also their weapon-grade plutonium. The present report incorporates the latest data on nuclear weapon dismantling and elaborates on the proposed agenda, but its basic recommendations differ little from the interim proposals.

This study was requested by the Office of the Deputy for Non-proliferation Policy, Office of the Under Secretary of Defense for Policy. The research was conducted within the International Security and Defense Strategy Program of RAND's National Defense Research Institute, a federally funded research and development center sponsored by the Office of the Secretary of Defense and the Joint Staff.

CONTENTS

FIGURES

TABLES

Since 1991, the governments of the United States and the former Soviet republics (FSRs), and the public worldwide, have become concerned about the rapid accumulation of weapon-usable fissile materials[1] from dismantled nuclear warheads. The fear is that mismanagement might result in some of the materials being refashioned into nuclear weapons, either by national or subnational groups or, if Russia or other nuclear republics revert to tyranny, by the republics themselves.

This study, however, found that countries including the United States have paid inadequate attention to an equally if not more serious potential danger on the civilian side. Current plans for civilian nuclear development worldwide call for the separation of more weapon-usable plutonium from spent fuel[2] by the year 2003 than from dismantled nuclear weapons. Another problem is the existence of commercial gas centrifuge and other sensitive[3] enrichment plants

---

[1]Weapon-usable fissile materials are defined as uranium with a fissile isotopic content of 20 percent or more and plutonium of any isotopic composition. Weapon-usable plutonium includes plutonium separated from the typical spent fuel of commercial nuclear reactors (reactor-grade plutonium) and plutonium from nuclear weapons (weapon-grade plutonium). On the other hand, plutonium before being separated from the intensely radioactive spent fuel is not considered as weapon-usable fissile material in this study.

[2]In this report, we define spent fuel as discharges from nuclear reactors before reprocessing to recover plutonium and uranium. Waste is defined as the aqueous streams containing dissolved spent fuel after plutonium and uranium have been recovered.

[3]For this report, sensitive civilian nuclear facilities are defined as those that can produce, separate, or handle weapon-usable fissile materials. These facilities include plants for plutonium reprocessing and fabrication, plutonium-fueled reactors, and at

in nonnuclear weapon states. Countries with separated plutonium or these enrichment facilities within their borders can, at will, produce materials for nuclear weapon use within days or weeks. No safeguard scheme, including that of the International Atomic Energy Agency (IAEA), can be effective if such sensitive materials and facilities are widely available in nonnuclear weapon states. The drafters of the Non-Proliferation Treaty (NPT) insisted at the outset that nuclear weapons and peaceful nuclear devices not be treated differently. We now argue that nuclear production facilities and sensitive civilian facilities also should not be treated differently; they should be confined to the currently declared five nuclear weapon states. A few exceptions, such as the Joyo and Monju breeders in Japan and the Urenco enrichment plants in the Netherlands, may need to be tolerated. Countries, however, should be encouraged to phase out their plutonium and enrichment activities or at least not to expand or build more.

It is critical that countries pay attention to the proliferation threat from the civilian side if they want to maximize the nonproliferation value of dismantling U.S. nuclear weapons and those of the FSRs. If countries ignore the civilian threat, they can compound the problem by making wrong choices in how to deal with military materials. As an example, some planners recommend burning weapon-grade plutonium in nonnuclear weapon states. This would actually encourage the civilian use of plutonium in those states.

This study recommends that the United States initiate and encourage countries to undertake a four-element program for managing civilian nuclear development. The elements are (1) terminating or drastically reducing both military and civilian plutonium activities worldwide, (2) prolonging the world's reliance on current once-through[4] and proliferation-resistant modes of nuclear plant operations, (3) focusing existing advanced nuclear reactor developmental programs on reactors (without plutonium reprocessing) that

---

least some of the plants for uranium enrichment. But a typical commercial nuclear reactor is not considered a sensitive nuclear facility, because it does not use weapon-usable fissile materials in its fuel and its produced plutonium is still embedded in intensely radioactive spent fuel.

[4]Almost all the current commercial nuclear reactors worldwide operate in the once-through mode, in which the plutonium and uranium in the spent fuel are not reused.

consume much less uranium and are more proliferation-resistant than current reactors, and (4) negotiating an international arrangement that allows sensitive civilian nuclear materials and facilities to exist and operate only in the five currently declared nuclear weapon states and that agrees on the sharing of benefits, if any, with nonnuclear weapon states. This four-element program would allow countries to use peaceful nuclear energy further into the future, with far less nuclear proliferation risk.

After delineating this four-element program, the study proposes complementary actions to deal with fissile materials from disman-tled nuclear weapons. Whatever is done to these military materials should meet two criteria. First, the actions should prevent FSRs as much as possible from ever fashioning these high-grade fissile materials back into nuclear weapons or selling them to nonnuclear countries or groups. Second, any actions taken should not hinder the international movement toward a proliferation-resistant future. Both criteria can be met by blending down the highly enriched uranium (HEU),[5] as both the Bush and the Clinton administrations have asked FSRs to do, and by purchasing weapon-grade plutonium from the FSRs. However, the United States is currently undecided on a course for plutonium. Storing and safeguarding plutonium in FSRs, as many planners propose, does not prevent FSRs from re-using the plutonium for bombs if FSRs politically change for the worse.

We propose in this study that the United States reduce its $8 billion to $12 billion[6] commitment by asking other countries to help pur-chase uranium blended down from FSR HEU. In this way, the United States can spend more on the purchase and management of FSR weapon-grade plutonium, which is much more difficult to make weapon-nonusable than HEU. Finally, the United States can give its own HEU and weapon-grade plutonium the same treatment as FSR

---

[5]In this report, we define highly enriched uranium as uranium with 90 percent or more fissile uranium isotopes.

[6]All dollar amounts in this report are in 1992 U.S. dollars unless specified otherwise. The $8 billion is calculated by assuming a 10 percent discount rate and the $12 billion is the undiscounted amount. The HEU value to FSRs is $6 billion to $9 billion correspondingly, because they will have to spend $2 billion to $3 billion to blend their HEU into low-enriched uranium.

materials, but allowance should be made for the differences between these countries' requirements for such materials (such as the need for HEU in U.S. naval reactors) and between their abilities to produce such materials quickly (such as the capability of FSR RBMK power plants to produce weapon-grade plutonium quickly after treaty abrogation).

RAPID ACCUMULATION OF WEAPON-USABLE MATERIALS

Preventing nuclear materials from falling into illegitimate hands has always been the most important technical element in international nuclear safeguards, and rightly so. If weapon-usable fissile materials became readily available commercially to both nuclear and nonnuclear weapon states, an effective control would be infeasible. Weapon-usable fissile materials could come from two sources: One is from dismantled nuclear weapons in the FSRs and United States. We estimate that over the next 10 years 200 tonnes of plutonium and 1,000 tonnes of HEU will be recovered from those dismantled weapons.

A second source is the plutonium reprocessed from the spent fuel of commercial power plants. We estimate that through the year 2003, 330 tonnes of reactor-grade plutonium will be separated from spent fuel. The diversion of even a tiny fraction of these materials will be enough to make many nuclear weapons. Only about 5 kg of weapon-grade plutonium or 15 kg of HEU are needed to make a primitive nuclear weapon in the kiloton range. Even reactor-grade plutonium is weapon-usable material—this was proved in a 1962 test in the United States.[7] The theoretical critical mass with reactor-grade plutonium is merely 7 kg.[8] Thus, the amount of reactor-grade plutonium needed for a kiloton-range bomb is merely 40 percent more than that needed for a weapon-grade plutonium bomb. By the year

---

[7] Letter from Wilbur A. Strauser, Chief, Weapons Branch, Division of Classification, Energy Research and Development Administration, to Richard Bowen, Division of International Security Affairs, August 4, 1977.

[8] We used the data on isotopic composition of reactor-grade plutonium given by Robert Selden, "Reactor Plutonium and Nuclear Explosives," Lawrence Livermore Laboratory, n.d., and interpolated the data in the critical mass table provided by Theodore Taylor, "Nuclear Safeguards," *Annual Review of Nuclear Science*, Vol. 25, 1975, p. 413.

2003, there will be enough surplus plutonium from dismantled nuclear weapons to make 40,000 primitive bombs (Figure S.1). Reactor-grade plutonium separated from civilian spent fuel will be sufficient to make 47,000 bombs. By 2010, although the amount of plutonium available from dismantled nuclear weapons is expected to stay about the same, plutonium separated from spent fuel will be sufficient to make 71,000 bombs.

Before discussing the U.S. policy toward fissile materials from FSR dismantled nuclear weapons, we discuss timely warning in nuclear safeguards and elaborate on a desirable path for the world's civilian nuclear development.

TIMELY WARNING IN NUCLEAR SAFEGUARDS

Effective safeguards do not merely detect the diversion of nuclear materials or facilities for making bombs. If the diverted materials or facilities are in such form that nuclear weapons can be made in days

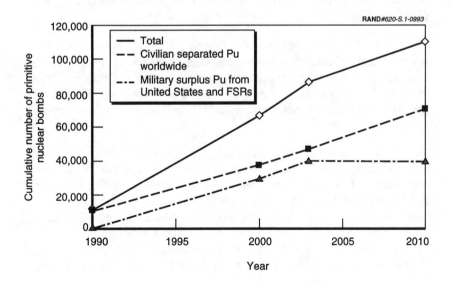

Figure S.1—Number of Primitive Nuclear Bombs That Can Be Made
from Separated Plutonium

or weeks, the United States and other countries would not have enough time to amass sufficient political and other pressures to prevent the completion of the bomb-making process. The warning time from detection to bomb production needs to be at least about a year. Notwithstanding the procedures and devices currently used by IAEA to safeguard sensitive facilities, it is an illusion that they can be effectively safeguarded according to any sound view of timely warning. These sensitive materials and facilities are just too difficult for any safeguard system to handle. The newer commercial enrichment plants in countries such as Japan, Germany, and the Netherlands are based on gas centrifuge, and converting them to the manufacture of weapon-usable uranium would take only days. The international community should also worry about dedicated sensitive military facilities such as the enrichment plant in Pakistan. As for plutonium, separated plutonium held in inventory could be diverted and reworked to make it weapon-ready in only days or weeks.

RECOMMENDED PATH FOR FUTURE CIVILIAN NUCLEAR DEVELOPMENT

There is no need to proliferate the enrichment facilities worldwide to meet countries' nuclear energy needs. Rather, enrichment facilities should be confined to current nuclear weapon states. With enrichment services available from several sources of different ideologies, a country need not be concerned about its supply being cut off.

Currently, civilian use of plutonium is not yet widespread. In fact, civilian nuclear power based solely on uranium is preferable to using plutonium. Plutonium use creates no economic benefits but much proliferation risk. Both thermal recycle[9] and plutonium-fueled fast reactors are not and will not be economically competitive with the current nuclear plants operating in the once-through mode. We estimate that thermal recycle will be uneconomical until the price of uranium-bearing yellowcake rises to \$100/lb U_3O_8. We project that

---

[9]Thermal recycle is defined as the operations of reprocessing plutonium and uranium from spent fuel and using plutonium-bearing mixed-oxide (MOX) fuel in thermal, the current type, nuclear power plants.

that price will not be reached until 50 years from now. We further project that fast reactors are not expected to be profitable until the yellowcake price reaches $220/lb U_3O_8 100 years from now. Even in an unlikely scenario extremely favorable to plutonium use, thermal recycle and fast reactors will not be profitable until the yellowcake price reaches $50/lb U_3O_8 and $140/lb U_3O_8, respectively. Adjusting for the situation in this scenario in which higher nuclear capacity growth leads to a faster rise in uranium price, we project that it will still take 30 years and 50 years, respectively, for thermal recycle and fast reactors to be economical. Both thermal recycle and plutonium-based fast reactors were planned worldwide during the early days of nuclear power, when projected nuclear capacity was routinely overestimated and uranium resources underestimated.

Countries now have enough time to explore a nuclear future that is more proliferation-resistant. In keeping with this goal, we recommend a four-element civilian nuclear program.

First, the United States, Canada, Sweden, and others who have indefinitely postponed plutonium use should urge other countries to terminate or slow their plutonium activities. One way to discourage plutonium activities is to not renew reprocessing contracts with the United Kingdom and France or to convert current or future reprocessing contracts into contracts to store or dispose of spent fuel.

If the United States cannot convince the United Kingdom and France to scale down their plutonium activities outright, a second-best option would be to encourage them to use weapon-grade plutonium from FSRs instead of reprocessing additional plutonium from spent fuel. Leaving plutonium in spent fuel is a practical and inexpensive way for all countries to save plutonium for unexpected future use. Countries should store spent fuel instead of separated plutonium.

This element will affect industrialized, nonnuclear weapon states such as Japan the most. Japan should not maintain all of its plutonium activities. After all, in the July 1993 Group-7 summit, Japan refused to endorse an indefinite extension of the NPT and wants to retain the option of developing nuclear weapons.[10] Although the

---

[10]Jim Mann and Leslie Helm, "Japan Shifts Its Stand on Ruling Out A-Bomb," *Los Angeles Times*, July 9, 1993, pp. A1 and A9.

government has reversed this position,[11] it or subsequent administrations can change its mind again. Japan should at least be urged to scale back its plan to use plutonium in 12 commercial reactors by the year 2005 to only two reactors and to cancel its planned construction of two plutonium-fueled demonstrators—the Demonstrator Advanced Thermal Reactor and the Demonstrator Fast Breeder Reactor. Neither will they need to construct additional supporting facilities for reprocessing and fabricating plutonium. Reducing to two reactors for thermal recycle plus the two breeder demonstrators—Joyo and Monju—should satisfy Japan's insistence on developing plutonium technology. (It may be willing to forgo development completely if the fourth element of our program, to be discussed below, is adopted.)

The second element in our program aims to extend the period during which countries can rely on current nuclear power plants, which operate in the once-through mode and are proliferation-resistant. Countries could pursue programs to improve uranium efficiency in current reactors, by high burnup, for example. The United States should initiate a joint effort with other countries to assess better the extraction costs and amounts of uranium resources, both the conventional and unconventional types, such as those from marine phosphates and seawater. A systematic evaluation will likely further enhance confidence that the earth has plenty of uranium to support the current types of once-through reactors well into the next century and beyond.

The third element in our proposed program is to shift the current emphasis on advanced reactor development programs worldwide to proliferation-resistant reactors. Reactor concepts have already been proposed in which uranium is consumed at a much lower rate—less than one-fifteenth of the uranium used in current nuclear power plants. In other words, if there is enough uranium to support any given level of nuclear capacity for 30 years, such new reactors, once fully deployed, could support the same capacity for the next 450 years, which is long by any planner's standard. Even if these new concepts proved not to work, the world would still have enough time

---

[11]Jacob Schlesinger, "Japan Supports Open Extension of Nuclear Treaty," *Wall Street Journal,* September 28, 1993.

to return to the traditional plutonium-bearing fuel cycles and reactors—thermal recycle and breeders.

Our fourth element is to establish an international arrangement for the contingency that plutonium use turns out necessary after all. Although we prefer to see countries eliminate all use of plutonium, some countries such as the United Kingdom and France have made a substantial financial commitment in plutonium and might not be willing to mothball their existing facilities or stop those under construction. If they continue plutonium activities, the related sensitive facilities, as well as those for uranium enrichment, should be confined within existing declared nuclear weapon states. Any exceptions should be eliminated when this fourth element is implemented. During the interim, exceptions should be made only rarely, where mothballing or moving existing plants to a nuclear weapon state or stopping projects well under construction will create a severe financial hardship. Some such plants in Japan and the Netherlands may qualify for exemption. No exception should be made for plants or upgrades still in the planning stage and in early stages of construction.

To placate nonnuclear weapon states, those nuclear weapon states participating in sensitive civilian activities should agree to share fully any benefits through energy credits and rebates. Such an agreement should assure nonnuclear weapon states that they can terminate their own programs on sensitive civilian activities without compromising much on their future security of supply and commercial competitiveness. Furthermore, nonnuclear weapon states should be free to conduct research, development, and production of non-sensitive components of sensitive systems. Such components include steam generators, heat exchangers, valves, temperature and other sensors, and gauges.

The Administration's aim to cut off the production of plutonium and HEU for nuclear weapons in all countries is worthy enough. We recommend that it be extended to include the cutoff of plutonium separation from power and research reactors. Moreover, the United States should propose that sensitive enrichment plants, such as the popular civilian centrifuge enrichment plants, be confined to the five declared nuclear weapon states. Below, we discuss policies toward FSR military fissile materials that are consistent with the goal of a desirable civilian nuclear future—one that is proliferation-resistant yet requires little economic sacrifice, if not actually bringing financial gains.

RECOMMENDED U.S. POLICY TOWARD HIGHLY ENRICHED URANIUM FROM FSRs

Blending HEU with natural or depleted uranium to produce low-enriched uranium (LEU) as soon as HEU comes out of the dismantled nuclear weapons essentially eliminates the weapon-usable form of uranium. Moreover, the HEU is highly valuable, because the value of the resulting LEU for reactor use is much higher than the blending cost. The 640 tonnes of HEU becoming surplus over the next 10 years in FSRs is worth $6 billion, and the 340 tonnes of surplus HEU in the United States is worth $3 billion.[12] The resulting LEU will meet about half the annual requirements for natural uranium and enrichment worldwide over the next decade. We recommend that the United States not release its military uranium, to soften any market disruption. Then the FSR military uranium alone will account for a smaller, but still sizable, 30 percent of the market. We further recommend that countries use blended-down uranium in their nuclear reactors and that they stockpile natural and low-enriched uranium as a means to absorb the excess supply and, as some countries are still worried about unexpected uranium shortage, to enhance security of supply. We also recommend that the United States encourage other countries to make purchases directly from FSRs or to repurchase what the United States has already bought. Such transactions are practical ways for countries to help FSRs. The United States should not have to shoulder all the financial burden of uranium purchases, especially when FSRs are likely to have twice as much HEU as many had thought only recently. Finally, the United States should favor conducting the blending operations in FSRs to create jobs for their defense and other workers.

---

[12]We assumed that the materials will become available in equal annual amounts throughout the next 10 years and that the annual discount rate is 10 percent. We estimated the undiscounted figures to be $9 billion and $4.5 billion, respectively. It has also been reported that the HEU in FSRs might be twice as much. Thus, the value would be doubled.

RECOMMENDED U.S. POLICY TOWARD WEAPON-GRADE PLUTONIUM FROM FSRs

We have examined five options for dealing with weapon-grade plutonium. The first is to use the plutonium as fuel in existing fast reactor demonstrators without reprocessing. The second option is to use it in light water reactors fueled with one-third or partial plutonium-bearing MOX without reprocessing. We call this option LWR (PM, w/o R). The third option is to use the plutonium in light water reactors fully fueled with MOX without reprocessing—LWR (FM, w/o R). The fourth is to store plutonium for, say, 20 years. The last option is to dispose of the plutonium by mixing it with waste or spent fuel when the waste or spent fuel is being prepared for final disposal. None of these options produces any commercial value for weapon-grade plutonium.

In the first three options, even if the weapon-grade plutonium is free, the extra cost in handling the highly radioactive and toxic plutonium outweighs the savings from using less uranium and enrichment. Using weapon-grade plutonium as fuel in fast reactors actually has a net cost of $18,000/kg; in LWR (PM), $7,600/kg; and in LWR (FM), $5,600/kg. The storage cost for 20 years is $3,800/kg. One way to dispose of plutonium is to mix it with waste or spent fuel being prepared for final disposal. The marginal cost for this approach is $1,000/kg. The U.S. repository, however, will not be ready for operation until 2010, and neither FSRs nor other countries have such repositories. Thus, to adopt the disposal option, interim plutonium storage cost must also be factored in. Even in the three options of using plutonium as fuel, plutonium storage cost might have to be paid for up to 10 years, because the reactors may not be ready for plutonium-bearing fuel immediately. Taking the storage cost into account, we find that the cost differences among the fueling options in LWRs and the store-now-and-dispose-later option are all between $4,000/kg and $10,000/kg. Although the difference in total cost for handling the FSR 110 tonnes of weapon-grade plutonium might amount to $660 million, that is not extremely large by national standards. The key policy factor should still be the proliferation risk in each option, not economics. On the other hand, since blending down HEU resolves the proliferation risk, economics becomes the key consideration for the HEU policy.

Even at $10,000/kg, the cost to eliminate the FSR 110 tonnes of weapon-grade plutonium would be $1.1 billion. This is not high relative to the potential risk of leaving it in the hands of FSRs. We recommend that, instead of charging FSRs $1.1 billion to eliminate their weapon-grade plutonium, the United States, alone or with some help from the United Kingdom and France, should buy it for, say, the same price. With the purchase, the primary objective of taking weapon-grade plutonium out of the unstable FSRs is accomplished. Whether the plutonium is stored or burned depends on what bargain we can strike with FSRs and our allies. If it is stored, the United States is the preferred location, but storage in the United Kingdom or France would be acceptable too. If it is burned, that could be accomplished in the already available plutonium-bearing fabrication facilities and nuclear power plants in the United Kingdom and France. But the United States should engage in a negotiation with them to reduce the amount of plutonium to be recovered from spent fuel. This way, weapon-grade plutonium could be eliminated while discouraging reprocessing. Thus, the United States should seek money from the United Kingdom and France to purchase FSR weapon-grade plutonium and/or should ask them to burn this plutonium without charging the United States or FSRs a fee.

Even after the best efforts of the United States and others, the FSRs might still refuse to let their weapon-grade plutonium leave the country, even though it has no economic value and will cost the FSRs money to manage. Then, a second option would be to dispose of it in the FSRs. Unfortunately, like other countries, the FSRs might not have a suitable repository for the next 10 to 20 years. Using weapon-grade plutonium as fuel in the FSRs also faces problems. For at least the next several years, FSRs will lack the needed fabrication facilities and appropriate nuclear reactors to eliminate the weapon-grade plutonium released from their dismantled weapons.

Nonnuclear weapon states such as Japan and Germany could help the FSRs build MOX fabrication plants or modify nuclear reactors to use plutonium-bearing fuel in the FSRs, but Japan and Germany (and other nonnuclear weapon states) should not conduct these activities in their own countries. Otherwise, other nonnuclear weapon states, such as North Korea and Iran, might be unwilling to forgo sensitive civilian activities or plans.

Whatever the nuclear weapon states do to dispose of their weapon-usable materials or to restrict their production, their activities should not encourage nonnuclear weapon states to start or continue sensitive civilian or military nuclear programs. For this reason, sensitive enrichment facilities should not be placed in nonnuclear weapon states and the world should not plunge into a plutonium economy. The economic benefits of plutonium use are distant and uncertain. The nuclear proliferation risks, however, are very real. Therefore, indefinite postponement of plutonium activities will not only save money but also make the world safer.

ACKNOWLEDGMENTS

We thank Gregory Jones for providing us with useful data and comments throughout the study. Bruno Augenstein, Bridger Mitchell, Richard Speier, and especially Henry Sokolski also provided us with many insightful comments on our draft. We also appreciate Dee Lemke's help in processing the report.

ABBREVIATIONS

AIF	Atomic Industrial Forum
ATR	advanced thermal reactor
CRBR	Clinch River Breeder Reactor
DoD	Department of Defense
FBR	fast breeder reactor
FMSR	fast mixed-spectrum reactor
FSR	former Soviet republic
GESMO	Generic Environmental Impact Statement for Mixed Oxide Fuels
GWe	gigawatt-electric
HEU	highly enriched uranium
HLW	high-level waste
HM	heavy metal
IAEA	International Atomic Energy Agency
JNFL	Japan Nuclear Fuel, Limited
LEU	low-enriched uranium
LMFBR	liquid metal fast breeder reactor
LMFR	liquid metal fast reactor
LWR	light water reactor
LWR (FM)	light water reactor (full MOX fuel)
LWR (FM, w/o R)	light water reactor (full MOX fuel, without reprocessing)
LWR (OT)	light water reactor (once through)
LWR (PM)	light water reactor (partial MOX fuel)
LWR (PM, w/o R)	light water reactor (partial MOX fuel, without reprocessing)
MOX	mixed oxide
MT	metric ton or tonne

MTHM	metric ton of heavy metal
MUF	material unaccounted for
NEA	Nuclear Energy Agency
NPT	Non-Proliferation Treaty
NRC	Nuclear Regulatory Commission
OECD	Organization for Economic Co-operation and Development
O&M	operating and maintenance
PFR	prototype fast reactor
PNL	Pacific Northwest Laboratory
ppm	parts per million
Pu	plutonium
Puf	Pu fissile
PWR	pressurized water reactor
RDT&E	research, development, test, and evaluation
START	Strategic Arms Reduction Treaty
SWU	separative work unit
UF_6	uranium hexafluoride
UO_2	uranium dioxide
U	uranium
WIPP	Waste Isolation Pilot Plant

Chapter One

INTRODUCTION

This report deals with the problem of rapidly accumulating weapon-usable fissile materials, defined as uranium with a fissile isotopic content of 20 percent or more, and plutonium of any isotopic composition (including both weapon-grade and reactor-grade plutonium).[1]

The problem has two parts. One has to do with the recovery of weapon-grade plutonium and highly enriched uranium (HEU)[2] from dismantled nuclear weapons over the next decade and beyond. The other part relates to the civilian side and the fact that an even greater

---

[1]The term "fissile" refers to nuclear materials that are fissionable by *both* slow (thermal) and fast neutrons. Fissile materials include U-235, U-233, Pu-239, and Pu-241. Materials such as U-238 and thorium-232 (Th-232), which can be converted into fissile materials, are called fertile materials. Note that Th-232, U-238, and all plutonium isotopes are fissionable by fast neutrons but not by thermal (slow) neutrons. They are not called fissile materials but may be called fissionable materials. The International Atomic Energy Agency (IAEA) and others have long used 20 percent enrichment as the cutoff point for "direct-use nuclear material" (IAEA, 1987, p.11). The Department of Defense uses the following terminology for plutonium. Super-grade plutonium contains 2 to 3 percent of Pu-240; weapon-grade, less than 7 percent; fuel-grade, 7–19 percent; and reactor-grade, 19 percent or greater (DoD, 1981, p. 104). Under the definition in this report, all grades of plutonium are weapon-usable, with a few exceptions. The most notable one is plutonium with 80 percent or more of Pu-238, which is exempted from IAEA safeguards (*The Structure and Content of Agreements Between the Agency and States Required in Connection with the Treaty on the Non-Proliferation of Nuclear Weapons,* p. 11). High concentration of Pu-238 might produce too much heat during its decay and melt the bomb materials. Also, plutonium before being separated from the intensively radioactive spent fuel is not considered as weapon-usable fissile material in this study.

[2]For this report, we define highly enriched uranium or weapon-grade uranium as uranium with 90 percent or more fissile uranium isotopes.

quantity of reactor-grade plutonium will be separated from spent fuel over the same period of time.[3] Two approaches have been advanced to prevent weapon-usable materials from falling into nuclear aspirants' hands, and these proposals could lead to quite opposite results.

Some planners believe that both sensitive nuclear materials and facilities, such as separated plutonium (Pu), plutonium reprocessing, and Pu/U (uranium) mixed-oxide (MOX) fabrication plants and gas centrifuge enrichment plants, can be safeguarded by the IAEA.[4] Thus, they advocate using military fissile materials in civilian power plants to stretch out the use of nuclear energy. This, they argue, will help convince nonnuclear weapon states that the nuclear states are sincere in their pledge to reduce their nuclear arsenals and to share with nonnuclear weapon states the benefits of nuclear power—an important goal, given that the Non Proliferation Treaty (NPT) is up for renewal in 1995.

Other planners, however, are more skeptical of IAEA's ability to provide adequate warning time for the world to react to violations. This concern existed before the Gulf War, and the Iraqi experience has merely reinforced it. In fact, the critical importance of warning time was recognized from the start. In 1946, the *Acheson-Lilienthal Report on the International Control of Atomic Energy* stated clearly that an effective safeguard system must provide adequate warning:

> It must be a plan that provides unambiguous and reliable danger signals if a nation takes steps that do or may indicate the beginning of atomic warfare. Those danger signals must flash early enough to

---

[3]In this report, we define spent fuel as discharges from nuclear reactors before reprocessing to recover plutonium and uranium. Waste is defined as the aqueous streams containing dissolved spent fuel after plutonium and uranium have been recovered. It should be noted that plutonium from commercial reactors is not always of reactor grade. They have occasionally been operated under conditions that yield plutonium of a better grade for making nuclear weapons.

[4]For this report, sensitive materials and weapon-usable materials are synonymous. Separated plutonium is plutonium separated from spent fuel. Further, sensitive civilian nuclear facilities are defined as those that can produce, separate, or handle weapon-usable fissile materials. But a typical commercial nuclear reactor is not considered a sensitive nuclear facility, because it does not use weapon-usable fissile materials in its fuel and the plutonium it produces is still embedded in intensely radioactive spent fuel.

leave time adequate to permit other nations—alone or in concert—to take appropriate action.[5]

The importance of warning time should not be forgotten. IAEA safeguards rely essentially on monitoring and accounting to detect illegal nuclear material diversion. Measurement uncertainties and the gradual accumulation of material unaccounted for may make it difficult to account accurately for diverted material if the amount of material under safeguards is huge and is frequently moved.[6] Moreover, IAEA (or any international body) has only limited intelligence capacity and ability to enforce the rules. No intelligence community can know all the major nuclear facilities and activities in closed societies, such as those of Iraq, North Korea, and Iran. Moreover, even after bomb-making and other sensitive facilities are discovered, IAEA and other international agencies are severely limited in their ability to neutralize the violation or to punish the offenders. Imperfect information is the reality and must be taken into account. Ineffective safeguards of sensitive materials and facilities are not the fault of IAEA. These sensitive items are just too difficult for any safeguard system to handle. Although North Korea has put its withdrawal from NPT on hold, its nuclear ambition will be difficult to stem, fueling these planners' skepticism. Therefore, these planners would prefer, instead of plunging into a plutonium economy,[7] a more cautious strategy that uses only HEU from dismantled nuclear weapons to stretch the fuel supply but delays the massive commercial introduction of separated plutonium. Once nonnuclear weapon states possess separated plutonium, any international agency would find it difficult to retrieve that plutonium. Further, even in the unlikely event that the nonnuclear state agreed to turn over the plutonium, the world could never be sure that a small amount was not secretly withheld. Because a

---

[5]Department of State (1946), p. 38.

[6]For difficulties of safeguarding plutonium reprocessing and fabrication plants, see, for example, Miller (1990).

[7]The world economy involving the commercial production, transportation, and use of separated plutonium.

nuclear bomb requires only about 7 kg of reactor-grade plutonium,[8] retaining a few bombs' worth could escape the material accounting system devised by the IAEA.

The world is at the fork of these two paths. Should countries speed up or delay civilian use of plutonium? Are there strategies that allow countries to postpone plutonium use now but still hedge against an uncertain future, in which plutonium might be needed in the civilian nuclear fuel cycle? In addition to examining these issues in this report, we analyze ways to manage, use, or dispose of fissile materials from dismantled nuclear weapons. How can the United States and the former Soviet republics (FSRs) keep such materials from falling into the wrong hands? How can the United States prevent the materials from being refashioned into nuclear weapons in the event that the Russian or other republics revert to tyranny?

The issues dealing with military fissile materials are new. Few had expected the rapid collapse of the communist empire and both sides' commitment to reduce nuclear arsenals so drastically and quickly. On the other hand, there were intense domestic and international discussions during the 1970s about the research, development, and demonstration of plutonium use and its implications for nuclear proliferation. Studies and debates about whether plutonium use should be encouraged or discouraged dropped significantly during the 1980s. Yet many changes affecting the future of plutonium use continued to occur during those years. The changes in both the military and civilian sides and the policy interdependency of dealing with military and civilian fissile materials make it important to analyze these issues together and now. We begin by pointing out the inherent physical difference between weapon-usable plutonium and weapon-usable uranium. This inherent difference leads to very different ways of dealing with plutonium and uranium.

---

[8]Whenever we mention an amount of plutonium or uranium without specifying "fissile," we refer to the mass of plutonium or uranium, including all isotopes. An amount of fissile plutonium or uranium (Puf or Uf) refers to the mass of fissile isotopes only.

WEAPON-USABLE URANIUM AND WEAPON-USABLE PLUTONIUM

Both plutonium and uranium can be used by first-timers to make primitive nuclear bombs.[9]   The United States and FSRs are, thus, concerned about the security of plutonium and HEU recovered from dismantled nuclear weapons. The essential difference between HEU and plutonium is that HEU can be made weapon-nonusable by blending it down into low-enriched uranium (LEU), whereas plutonium mixed with other elements can easily be made weapon-usable again.

HEU can be diluted by the widely available U-238.[10]   Once diluted, the material is no longer weapon-usable. Although the critical mass of weapon-grade uranium is about 17 kg,[11] diluting it with U-238 down to 20 percent fissile content will raise the critical mass of uranium to 250 kg, as shown in Figure 1.1, an increase of a factor of 15.[12] Adding in other necessary components of a primitive nuclear bomb such as the reflector and high explosives would make a bomb using uranium with less than 20 percent of fissile content very heavy, and it would be impractical to develop a survivable delivery system for it. Therefore, the low-enriched or blended-down uranium would have to be re-enriched. But enrichment facilities are time-consuming and difficult to build. As long as the international nuclear control agencies do not allow sensitive enrichment facilities[13] to be built in non-nuclear countries, HEU can be made safe by dilution.

---

[9]The first nuclear bomb, the Little Boy, was a uranium bomb; the second one, the Fat Man, used plutonium.

[10]Natural uranium ore contains almost 99.3 percent U-238 and only 0.7 percent U-235.

[11]The U.S. HEU used in nuclear weapons is commonly known as oralloy and is enriched to 93.5 percent (Roser, 1983, p. 4979). Taylor calculated that, at 93.5 percent enrichment, the critical mass of uranium is 17 kg, assuming a sphere of metallic uranium surrounded by a 15 cm thick reflector of natural uranium (Taylor, 1975, p. 412).

[12]To help readers trace and replicate estimates throughout this report, we sometimes present numbers, especially during intermediate steps, beyond their significant figures. Numbers in the final results and recommendations, however, are rounded.

[13]We define a sensitive enrichment facility as one that can be reconfigured to produce weapon-usable uranium (i.e., with 20 percent or more fissile uranium) within a year. A gas centrifuge enrichment plant is a sensitive enrichment facility, because the

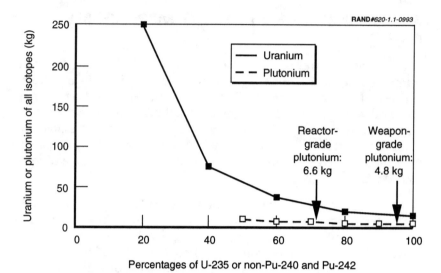

SOURCE: A plot based on data in Taylor (1977), pp. 412–413.

Figure 1.1—Critical Masses of Uranium and Plutonium

From a security standpoint, plutonium is very different from uranium. It is difficult to make plutonium unusable for nuclear bombs. Uranium is relatively abundant in the earth's crust and ocean waters, but they contain only traces of plutonium. There is no ready supply of natural plutonium ore for an isotope like U-238 to dilute weapon-grade plutonium. Nor does mixing weapon-grade plutonium with plutonium separated or extracted from civilian spent fuel make the blended plutonium unusable for weapons: All plutonium isotopes are fissionable by fast neutrons, which are the relevant neutrons in nuclear weapon design.[14] In fact, the critical mass of reactor-grade plutonium is only 40 percent more than that of weapon-grade plu-

---

reconfiguration takes only days. If nonnuclear nations are allowed to build sensitive enrichment facilities, the difference between uranium and plutonium will be less significant.

[14]Typical weapon-grade plutonium has 93.5 percent Pu-239, 6 percent Pu-240, and 0.5 percent Pu-241. Typical reactor-grade plutonium has 1.5 percent Pu-238, 58 percent Pu-239, 24 percent Pu-240, 11.5 percent Pu-241, and 5 percent Pu-242 (Selden, n.d.).

tonium (Figure 1.1).[15] This is not merely conjecture. In 1977, the
then Energy Research and Development Administration declassified
the information that the United States in 1962 had tested a nuclear
bomb using reactor-grade plutonium, which resulted in "a nuclear
yield."[16] Diluting plutonium with U-238 would make the mixture
unusable for weapons. But nuclear aspirants can separate the
plutonium from U-238 chemically, which is much easier to do than
isotopic enrichment.

That plutonium cannot be made weapon-unusable while uranium
can has far-reaching implications for the ways these two types of
fissile materials are treated and managed.

ORGANIZATION OF THIS REPORT

In Chapter Two, we first estimate the amount of plutonium and HEU
that will become available from FSR and U.S. nuclear arsenals. Then,
we estimate the amount of separated plutonium that can be
obtained from spent fuel worldwide. In Chapter Three, we examine
plans to use plutonium in the civilian sector worldwide. In Chapter
Four, we analyze the options in using and managing plutonium,
especially the weapon-grade plutonium from dismantled weapons,
with particular attention to the economic and political benefits or
costs in each option. In Chapter Five, we deal with HEU. In Chapter
Six, we summarize our recommendations on the paths that civilian
nuclear development worldwide should take and on how to manage
the plutonium and HEU recovered from nuclear weapons so that the
world can reap the benefits of nuclear power without incurring the
risks of undue proliferation.

---

[15]The critical mass of weapon-grade plutonium with a thick natural uranium reflector
is 4.8 kg and that of reactor-grade plutonium is 6.6 kg, or an increase of about 40
percent. We derived these numbers by using isotopic composition data from Selden
(n.d.) and by interpolating data from Taylor (1975), p. 413.

[16]Strauser (1977). See also Wohlstetter (1977), p. 21; Gillette (1977a), p. 1; and Gillette
(1977b), p. 3.

Chapter Two

QUANTITIES OF WEAPON-USABLE PLUTONIUM AND URANIUM

In this chapter, we estimate the amounts of weapon-usable plutonium and uranium from both the military and civilian sectors worldwide. These estimates will be used in later chapters to analyze the costs and market effects of various options to manage fissile material. We also discuss how much fuel for nuclear power plants can be substituted by weapon-usable plutonium and how much by weapon-usable uranium. If only a small amount of fuel is available from these weapon-usable materials, disposing of them instead of using them will not be much worse from the perspective of resource conservation. On the other hand, if large amounts of fuel are involved, the conversion and release schedule will require particular attention so that the market effect is not too disruptive. This chapter also shows the number of primitive bombs that can be made from these fissile materials, to indicate the magnitude of the problem. Finally, we discuss how weapon-usable materials might be diverted and used and, in subsequent chapters, how to handle these worrisome situations.

FISSILE MATERIALS FROM WEAPONS

In 1991, the United States had about 19,000 strategic and tactical warheads and the FSRs had 32,000, including 5,000 old warheads in storage.[1] With a reduction of strategic warheads under the Strategic

---

[1]Albright et al. (1993), pp. 36 and 40. Hereinafter cited as *World Inventory* or Albright et al. (1993). We will use their central estimates in this report (see their book for

9

Arms Reduction Treaty (START 1) and of tactical weapons under various initiatives, each side, by 1991, had committed to reduce its total number of nuclear warheads to about 10,000. START 2, signed by Presidents Bush and Yeltsin in January 1993, calls for a reduction to 3,500 U.S. strategic warheads and to 3,000 FSR warheads by the year 2003. The number of tactical nuclear warheads to be kept by each side depends on future unilateral or bilateral actions. We assume that the United States will keep 1,500 tactical weapons and the FSRs will keep 2,000. Thus, each side will have 5,000 strategic and tactical warheads by the year 2003 (Figure 2.1).

The amount of fissile material that will become available from nuclear weapons depends on the schedule of dismantling. Currently, both sides eliminate about the same number of warheads (2,000 to

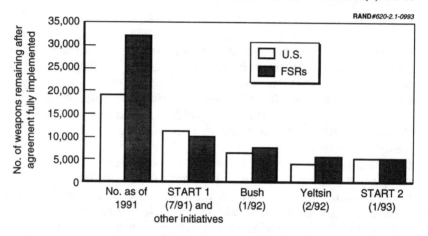

SOURCES: *Armed Forces Journal International*, (1992), p. 28; and Albright et al. (1993); pp. 36 and 40.

Figure 2.1—Recent Agreements and Initiatives on Nuclear Weapon Reduction

margins of error). Much of our research and many calculations were performed before the appearance of *World Inventory*. After we compared and checked their data, we were satisfied that their technical data form a credible and consistent basis for analyses. We have recalculated some of our numbers using their figures as input. Their figures are cited throughout this report. It should be noted that their study was not focused on the assembly of economic data, which we have developed from other sources as cited.

2,500) a year.[2] Since the United States has fewer nuclear weapons to eliminate, dismantling at the same rate will place the United States ahead of the FSRs, which we do not recommend. Unless the FSRs speed up their dismantling rate, the United States should slow its pace. In this report, we assume that the latter is the case. Hence, we estimated that the United States will destroy only 1,170 warheads a year.[3] On the other hand, FSRs will dismantle 2,250 per year, which reflects the current rate.[4]

Using the above dismantling schedule, we can determine the amount of weapon-usable materials remaining in nuclear warheads and military inventories and surplus materials by 2003[5] (Table 2.1). Since the United States currently plans to continue using HEU as a fuel for naval reactors, we assume that 48 tonnes will have been consumed by 2003 and that a 20-year inventory of 80 tonnes will be held. By 2003, there will be a surplus of 200 tonnes of plutonium and 1,000 tonnes of HEU from both sides. If these materials were released for civilian uses, they could fuel all current nuclear power plants worldwide for about 6 years.[6] Table 2.2 breaks down the contribution of plutonium and HEU from the United States and the FSRs.

Four points can be made from the data.

First, HEU is much more important than plutonium in fueling current commercial nuclear power plants. Plutonium would contribute only 13 percent of the energy contained in the surplus military fissile

---

[2]Berkhout et. al., (1992), p. 30.

[3]We assume that the United States will dismantle 14,000 weapons (i.e., 19,000 minus 5,000) at a constant rate over the 12-year period 1991–2003 (i.e., 1,170 warheads a year).

[4]We assume that the FSRs will dismantle 27,000 weapons (i.e., 32,000 minus 5,000) at a constant rate over the 12-year period 1991–2003 (i.e., 2,250 warheads a year).

[5]Materials in inventory are those that have not been placed into warheads. The United States uses weapon-grade uranium to fuel its naval reactors; the FSRs do not. In this estimate, we used the historical average demand of 4 tonnes a year or 48 tonnes during 1991-2003. (The figure of 4 tonnes a year came from *World Inventory*, p. 53. We also used the basic data in *World Inventory* for our estimates and projections in Table 2.1.)

[6]We based these estimates on current worldwide nuclear capacity without taking into account the projected small 1.8 percent per year growth in nuclear capacity (see Figure 3.4, below). Otherwise, the number of years would be somewhat lower.

Table 2.1

Tonnes of Weapon-Usable Fissile Materials in U.S. and FSR Warheads and Military Stockpiles

		1991		2003		Surplus	
		Pu	HEU	Pu	HEU	Pu	HEU
United States	Warheads	67	285	17.5	75	49.5	210
	Military stockpiles	43	265	2	88	41	129[a]
FSRs	Warheads	107	505	17.5	75	89.5	430
	Military stockpiles	21	215	2	8	19	207
	Total	238	1,270	39	246	199	976[a]

[a]Assumes that 48 tonnes of HEU have been consumed in naval reactors.

Table 2.2

Length of Time Current Nuclear Power Plants Worldwide Could Be Fueled with Military Weapon-Usable Surplus Materials

	FSR Materials	U.S. Materials	Total
Plutonium	0.4 yr	0.3 yr	0.7 yr
Uranium	3.2 yr	1.7 yr	4.9 yr
Plutonium as a percentage of sum	11	15	13

materials. On the other hand, plutonium would be more useful for starting fast reactors in the future.

Second, since utilities have signed long-term contracts for a sizable portion of their requirements in yellowcake[7] and enrichment, releasing all the fissile materials from dismantled nuclear weapons over the next 10 years could significantly disrupt the market, which would

---

[7]Yellowcake is the common name for uranium ore concentrates and contains about 85 percent U_3O_8. In this report, we follow the customary convention of quoting yellowcake price in units of dollars per pound of U_3O_8, not per pound of yellowcake. In other words, unless specified otherwise, $/lb of yellowcake means $/lb U_3O_8 of yellowcake. Similarly, the weight of yellowcake refers to the weight of U_3O_8 in it, not the gross weight of yellowcake.

be inadvisable, because countries consider these materials and services to be of strategic importance. A sharp reduction in demand for yellowcake and enrichment over 10 years will likely result in layoffs and facility closings. Then, countries would be concerned that existing materials and services might be inadequate to meet any sudden surge in demand. Thus, the amount of weapon materials released to the commercial market should be reduced. Also, countries such as Japan and Germany should be encouraged to use FSR blended-down, enriched uranium in their power plants and to stockpile natural uranium. Stockpiling helps not only to increase demand but also to assure a supply in the event of an unexpected surge in demand, as these countries want that assurance.

Third, if the United States does not release its plutonium and HEU over the next 10 years, the market disruption will be much less. FSR military plutonium and HEU surplus released to the market will amount to 36 percent of the current uranium demand over a 10-year period.

Fourth, if plutonium is not allowed to be used in current reactors, the FSR blended-down HEU will amount to an even more manageable 32 percent of demand over the next decade. (Additional analysis of market disruption will be given in Chapter Five.)

Table 2.3 shows the number of primitive bombs that can be made with these surplus nuclear materials.[8] The HEU from dismantled

Table 2.3

Number of Primitive Bombs That Can Be Made from Surplus Weapon-Usable Materials by 2003

	FSRs	United States	Total
Plutonium	21,700	18,100	39,800
Uranium	42,500	22,600	65,100
Total	64,200	40,700	104,900

[8]We assume that 5 kg of weapon-grade plutonium or 15 kg of HEU are required to make a primitive bomb in the kiloton range. See, for example, Wohlstetter et al. (1977), p. 34. We further assume that 7 kg of reactor-grade plutonium would be required for a kiloton-range bomb.

nuclear weapons could create 65,000 bombs. An ideal nonproliferation solution would be to immediately blend down the HEU as soon
as it is recovered from dismantled nuclear weapons. But plutonium,
which could produce 40,000 primitive bombs, cannot be dealt with
in the same way and must be safeguarded or eliminated. We will examine this issue after discussing plutonium from civilian operations.

SEPARATED PLUTONIUM FROM SPENT FUEL

There is a second source of plutonium, in addition to dismantled
nuclear weapons. Spent fuel from current and past civilian nuclear
power operations contains a huge amount of plutonium. At the end
of 1990, there were 532 tonnes of plutonium in spent fuel worldwide.[9] The intense radiation in the spent fuel, however, would make
it difficult for any terrorist group to extract plutonium from it.[10] The
difficulty of diversion at the national level would depend on indige-
nously available facilities. If nonnuclear weapon states are allowed
to build and operate reprocessing plants, they can extract plutonium
within days after they illegally divert spent fuel from safeguarded
facilities. In fact, they could merely seize the inventory of separated
plutonium from their own reprocessing or fabrication plants. The
MOX fuel for the annual reload of a typical 1 gigawatt-electric light
water reactor (LWR) would contain enough separated plutonium for
about 70 nuclear bombs.[11]

Those concerned with international nuclear nonproliferation can
take one of two approaches. One approach is to try to place all nuclear facilities, at least those in nonnuclear weapon states, under
safeguards. The proponents of this scheme consider such safeguards
adequate. We think otherwise. If nonnuclear countries are allowed
to have reprocessing plants or, even worse, to save separated plutonium for future civilian use, even full-scope safeguards could not

---

[9]*World Inventory,* p. 199.

[10]The terms "extract," "separate," and "recover" are used interchangeably in this
report.

[11]We assume that one-third of the fuel will be MOX as traditionally planned.
Countries have also considered reactors that use 100 percent MOX fuel. The
plutonium in such a reactor will contain enough material for three times as many
bombs.

provide timely warning after a country decided to abrogate its safe-guard agreement.

The other approach is for nonnuclear countries to forgo reprocessing the spent fuel and storing separated plutonium. Then, if a country seizes spent fuel, it still has to spend 18 to 24 months to build the necessary reprocessing facility to extract the plutonium.[12] In such a scenario, timely warning can be obtained.

By the end of 1990, France, the United Kingdom, and others had sep-arated about 122 tonnes of plutonium from spent fuel, 50 tonnes of which had been used in fast reactor demonstrators and in thermal recycles,[13] leaving about 72 tonnes. By 2000, 188 additional tonnes of plutonium will be separated; by 2010, another 236 tonnes.[14] Thus, the amount of plutonium separated from spent fuel and available for civilian use will increase from 72 tonnes to 260 tonnes by 2000 and 496 tonnes by 2010. We will discuss the implications of this rapid growth in the next chapter.

SEPARATED PLUTONIUM FROM WEAPONS

Plutonium from dismantled nuclear weapons will amount to about 149 tonnes by the year 2000 and to about 199 tonnes by 2003.[15] Fig-

---

[12]Wohlstetter (1976), pp. 155–156. We further assume that sensitive enrichment facilities are also denied to nonnuclear states.

[13]Thermal recycle is defined as the operations of reprocessing plutonium and uranium from spent fuel and using plutonium-bearing MOX fuel in thermal or current types of nuclear power plants. A reactor can be partially or fully loaded with plutonium-bearing fuel. We use two specific designs: light water reactor with one-third or partial MOX fuel (LWR (PM)) and light water reactor with 100 percent or full MOX fuel (LWR (FM)). But for calculations and discussions about thermal recycle in this report, we use the LWR (PM) as the reference, because this is the most common mode being used or to be used in countries with thermal recycle plans. We also introduce the term "plutonium burners," which use plutonium-bearing fuel but not plutonium reprocessing. We define three plutonium burners: LWR (PM) without reprocessing (LWR (PM, w/o R)); LWR (FM) without reprocessing (LWR (FM, w/o R)), and a fast reactor without reprocessing (FR (w/o R)).

[14]*World Inventory*, p. 206.

[15]See Table 2.1 for our estimate of plutonium available by the year 2003. The estimate for the year 2000 assumed that the amount of surplus becomes available for civilian use at a constant rate from now until 2003, the year by which START 2 requires that scheduled nuclear dismantling be completed.

ure 2.2 shows that the combined amount of plutonium from spent fuel and nuclear weapons rises rapidly. Over the next decade, the amount of separated plutonium will increase by a factor of seven. Safeguarding this much plutonium, which is sufficient for 87,000 primitive bombs, presents two problems of particular concern.

The first problem relates to the FSRs. By 2003, they will have released 110 tonnes of plutonium from dismantled nuclear weapons. If these materials are kept within the FSRs and if Russia or other republics revert to a totalitarian society, these materials can be quickly fashioned back into nuclear weapons. IAEA or other safeguards of plutonium storage facilities within the FSRs can do little to reduce this danger of national confiscation. Therefore, the United States should attempt to get the weapon-grade plutonium out of the FSRs, even if the United States has to pay for it. Russia will likely resist the U.S. proposal initially. But if the United States is serious and persistent and if the offer is attractive, Russia may accept. It is more likely that the non-Russian republics, such as Ukraine, would accept the U.S. offer, because they would probably prefer to have the plutonium in U.S. rather than Russian hands. They worry more that some day the Russians, not the West, would attack them.

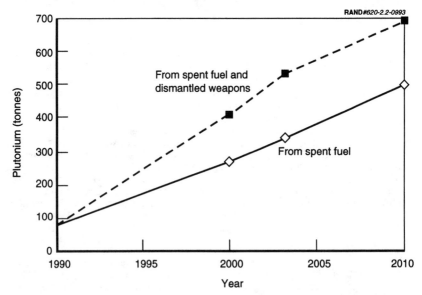

Figure 2.2—Separated Plutonium Potentially Available for Civilian Use

Our discussions with environmental and other public interest groups revealed that, in spite of environmental and political concerns, most are in favor of our proposal of plutonium purchase. We also argue that because the United States has to prepare to store plutonium from its own dismantled nuclear weapons, there will be economies of scale for storing both U.S. and FSR plutonium in the same place, especially because these facilities will have high fixed costs for safety and security. Finally, the West is providing billions of dollars in financial and other aid to the FSRs and is asking little in return. The West is justifiably concerned about 110 tonnes of weapon-grade plutonium in a highly unstable country (Table 2.1).

The Russians may object to this proposal, at least initially, even though they are willing to sell their HEU to the United States. They may view HEU and plutonium differently. First, they may have more HEU than the West knows. Selling the United States 500 tonnes of HEU could still leave them with as much as 700 tonnes.[16] Moreover, they could argue that HEU can be blended down to LEU and become nonweapon-usable. They may worry that the United States might save FSR plutonium indefinitely, retaining the option to some day make more weapons. If this concern jeopardizes the purchase proposal, the United States could consider storing FSR plutonium and perhaps even U.S. plutonium in a third country such as the United Kingdom. The United Kingdom is a good choice, because it already has an active reprocessing program and has made provision to store its own separated plutonium while it is waiting to be used in civilian reactors. The FSRs might still see this arrangement as advantageous to the United States, because the United Kingdom is a close U.S. ally. A last alternative would be to place both sides' military plutonium in a mutually acceptable country.

The United States should try to purchase all of the FSR weapon-grade plutonium, but getting only a portion of it remains an attrac-

---

[16]The amount of HEU in FSRs is highly uncertain, as reflected in the estimates of Albright et al. (1993). They based their estimates on enrichment capacity and presented a range of 520 to 920 tonnes of weapon-grade uranium in FSRs (*World Inventory*, pp. 57 and 60). Estimates from governmental and nongovernmental sources also reflect a wide range of uncertainties. For example, Steyn and Meade (1992), slide 1, use a figure as high as 1,100 tonnes in their analysis. In fact, Viktor N. Mikhailov, head of the Russian Ministry of Atomic Energy, said that the Russian inventory of HEU is more than 1,200 tonnes (Broad, 1993, p. 1).

tive option. After all, the United States and the FSRs always consider the number of nuclear weapons on each side important in their strategic calculation. That FSRs cannot quickly reassemble a large number of nuclear weapons will certainly reduce the cost to the United States of preparing a responsive posture during peacetime.

Some might argue that purchasing even all the military plutonium from FSRs would help the matter little, since FSRs can reprocess spent fuel to obtain plenty of plutonium. This argument is valid for nonnuclear weapon states but not for FSRs. A nonnuclear weapon state could thereby equip itself with a dozen or so nuclear weapons. In the case of FSRs, one is much less worried that FSRs will get an additional dozen or so nuclear weapons. Instead, the main concern is that they will reuse dismantled materials to reconstruct thousands of nuclear weapons quickly. Nonnuclear weapon states, when they cannot obtain weapon-grade plutonium to make primitive nuclear weapons, will settle for reactor-grade and can design appropriate delivery vehicles from scratch. Both the bombs and the delivery vehicles will be larger but still practical, as the critical mass is only 40 percent larger. These primitive bombs could still be delivered by aircraft and perhaps even missiles.[17] On the other hand, the FSRs have already designed their delivery vehicles. If they have to use reactor-grade plutonium for thousands of weapons, they will have to test and certify these new reactor-grade plutonium bombs and perhaps modify or redesign some of their delivery platforms—a costly and time-consuming operation. These tasks will make nuclear rearming less likely. Even if FSRs decided to rearm, the long warning time would allow an easier and less costly U.S. response.

Some FSR government officials or workers, facing dire economic conditions, may be tempted to steal plutonium from the storage facilities and sell it to nuclear-aspiring nations. Even if the plutonium is under IAEA safeguards, the chance of diverting plutonium from a storage site in an FSR is greater than in countries such as the United States, the United Kingdom, and France—another reason in favor of taking military plutonium out of FSRs.

---

[17]The Chinese CSS-2 missiles, which have been sold to Saudi Arabia, can carry a payload of 2,000 kg. Scud B can carry only 1,000 kg, but it could be modified to carry heavier payloads in the near future.

The United States should explore other options in case FSRs refuse to sell their weapon-grade plutonium. An equally satisfying solution would seem to be for the FSRs to dispose of their military plutonium by mixing it with waste produced when the military plutonium was extracted in the first place. The problem is that the FSRs do not have any operational repository, and it is uncertain when one will be available.[18] The military plutonium may end up being stored in FSRs indefinitely, remaining subject to the risk of being reused in weapons. For this reason, we consider purchasing FSR plutonium a better option than plutonium disposal within FSRs, because it might be stored but not disposed of for many years.

Burning plutonium in current commercial nuclear power plants, fast reactor demonstrators, and specially designed plutonium burners are the other options (discussed in Chapter Four).

The second situation of concern is the existence of separated plutonium in nations such as North Korea. The North Koreans could obtain a large amount of such materials through plutonium reprocessing from civilian nuclear activities. That North Korea has suspended its withdrawal from NPT does not eliminate the danger that it might some day seize plutonium from its reprocessing facilities. IAEA safeguards can provide timely warning only if sensitive civilian nuclear facilities and activities are not allowed. Simply prohibiting such activities in only currently disruptive states would be extremely difficult. The drastic political turmoil in Iran since the mid-1970s revealed that a country can become disruptive unexpectedly and quickly. In any case, the international community will have trouble agreeing on which states are destabilizing. A ban applied to all countries—whether nuclear or not and whether stabilizing or not—would have a much better chance of success than a discriminatory ban. But for a total ban to be practical, the economic sacrifice of countries giving up those sensitive civilian activities would have to be small or nonexistent. This makes apparent the importance of estimating the economic costs and benefits of sensitive civilian nuclear activities, a topic covered in Chapter Three.

---

[18]Repository is defined as a permanent disposal site.

PLUTONIUM USE WORLDWIDE IN CIVILIAN NUCLEAR FUEL CYCLES

This chapter deals with the status, plans, and economics of plutonium use in civilian nuclear power worldwide. At the end of the chapter, we present our recommendations about civilian nuclear policy, including future plutonium use. In Chapters One and Two, we showed that sensitive enrichment plants such as those based on gas centrifuge technology are now available in several nonnuclear weapon states and that 300 tonnes of plutonium will be separated from spent fuel by the year 2003 and a lot more in subsequent years. We believe that these plants and materials pose a long-term serious proliferation threat. It is critical that countries pay attention to this threat if they want to maximize the nonproliferation value of dismantling U.S. and FSR nuclear weapons. If countries are inattentive to the civilian threat, they can compound the problem by making wrong choices in how to deal with military materials.

Moreover, if countries focus, as they should, on the civilian threat first, it will lead to a course of actions that will also help guide countries in deciding what to do with dismantled nuclear materials from U.S. and FSR arsenals.

This chapter begins with a short history of plutonium use policy in key countries. The past momentum in pushing for plutonium use in many countries has been slowed down by high cost and concerns about proliferation, and also by the slow growth in civilian nuclear power. Therefore, there is a much better chance now than before of slowing plutonium use.

The chapter also introduces a model for calculating electricity cost generated by various nuclear systems or modes at a given yellowcake price. As more nuclear power plants are built, more uranium resources will be depleted, resulting in two possible situations. First, more uranium could be found, and the uranium price could stay unchanged. Second, although more uranium might be found, it might be more expensive to extract. Then, the uranium price could rise. Plutonium, being a substitute for uranium, would become more attractive as uranium prices rose. To place plutonium use in better light, we assume that the uranium price will rise over time. The uranium price at which plutonium use becomes economical is called the uranium breakeven price. In this chapter, the breakeven prices for thermal recycle and breeders will be calculated. Then, we estimate the uranium supply curve. By comparing it with the uranium demand, we determine the uranium price over time and also when thermal recycle and breeders will become economical. If these dates are distant, there would be time to develop other, more proliferation-resistant systems. Also, the actions that the United States and FSRs take in dealing with their military materials should conform with the civilian nuclear future that the world prefers.

We will quickly sketch the current status of civilian nuclear power as a way to introduce the basic nuclear systems and mode of operations. Current commercial nuclear reactors are thermal reactors. Most operate in the once-through mode (i.e., there is no reuse of recovered plutonium or uranium from spent fuel). Although countries have separated a substantial amount of plutonium from spent fuel, much of it has been stored or used in fueling the initial cores and reloads of the breeder demonstrators scattered in various countries. Only a small fraction has been used in thermal reactors as demonstrators of thermal recycle.[1]

PAST AND CURRENT PLUTONIUM USE POLICY

In the early days of the nuclear age, the Manhattan Project scientists thought that the earth's crust contained only 18,000 tonnes of eco-

---

[1] See the Glossary for a more detailed discussion of various reactors and technologies.

nomically recoverable yellowcake.[2] That amount could support only about four current-size nuclear power plants through their 30-year lifetime. Because there would be insufficient uranium for the United States, let alone for the whole world, to reap the benefits of thermal nuclear reactors, the atomic scientists' solution was the breeder reactor, which, while generating electricity, could convert the much more abundant fertile U-238 (as compared with the fissile U-235) into fissile Pu-239.[3]

The early estimate of uranium resources was far too pessimistic. Yet, even as much more uranium was discovered, many nuclear planners remained reluctant to seek more proliferation-resistant alternatives to breeders. Who could resist the clever idea of generating electricity while producing more fissile materials than consumed? The need for plutonium in breeder startup is a key reason for reprocessing spent fuel. During the 1950s and 1960s, using plutonium in thermal re-actors, in addition to breeders, also was featured prominently in countries' civilian nuclear developmental plans.[4]

These plans encountered two obstacles. They ran afoul of nuclear nonproliferation and economics. Full-scale reprocessing and breeder development will involve such a large amount of separated plutonium that fool-proof protection against diversion becomes difficult or even impossible. Indeed, India's explosion of a "peaceful" nuclear device in 1974 had an awakening effect on nonproliferation policies worldwide. Further, a number of studies showed that the use of plutonium in civilian nuclear plants will actually increase the cost of nuclear electricity, because the cost of plutonium handling outweighs the fuel savings and, in the case of breeders, the capital cost is higher.[5] Influenced by the Indian

---

[2]The figure is mentioned in Wohlstetter (1976), p. 15ff.

[3]Uranium ore contains only 0.71 percent fissile U-235 and 99.28 percent of the fertile U-238; thus, U-238 is 140 times as abundant as U-235.

[4]The current commercial reactors are thermal reactors, because thermal (slow) neutrons are used to trigger fissions. Breeders are fast reactors that breed more fuel than consumed. In this report, we use the terms breeders and fast reactors interchangeably, although when the breeding requirement is important, we tend to use the term breeder.

[5]For studies from that time, see Cochran (1974), Chow (1975), and Wohlstetter et al. (1975).

explosion and the findings of a major study, President Ford, at the end of his administration, announced that the United States would indefinitely postpone thermal recycling.[6] President Carter reaffirmed that position.[7] Further, his administration took a position against the Clinch River Breeder Reactor (CRBR) Demonstration Project. The construction of CRBR, however, was not terminated until the Reagan administration, because the Senate Majority Leader, Howard Baker, and many other congressional leaders continued to support the breeder program during the Carter administration.[8] By the 1980s, the United States, Canada, and Sweden decided to postpone indefinitely reprocessing and breeders, whereas France, the United Kingdom, the former Soviet Union, Germany, and Japan kept to their original nuclear development plans.

Today, nuclear demand worldwide is drastically lower (to be discussed below). Moreover, the breeder program in Germany has been canceled, and the U.K. government will withdraw its financial support to the Prototype Fast Reactor (PFR) in 1994. Even in France, the most ardent supporter of breeder development, the Superphénix has been out of commission since 1989 and needs to be relicensed before resuming operations. The FSR breeder program has also been scaled back. The plans to build four plutonium fast reactors in the South Urals have been changed; now only one will be built. Japan is the only industrialized country with its breeder program more or less unscathed so far.

As to reprocessing, France and the United Kingdom are the two major advocates. Also, the FSRs have a sizable reprocessing capability of 200 to 300 tonnes of spent fuel per year. Japan and India each

---

[6]Wohlstetter et al. (1975). Carl Walske, president of the Atomic Industrial Forum, said, in discussing the change in U.S. policy at that time, that "the most significant single event . . . was the appearance in December 1975 of [the study] for the U.S. Arms Control and Disarmament Agency entitled, 'Moving Toward Life in a Nuclear Armed Crowd?'" Quoted by Fred Iklé in the Foreword to Wohlstetter et al. (1977).

[7]The study by the Ford-Mitre Nuclear Energy Policy Group published at the beginning of the Carter administration was also highly influential (Ford-Mitre, 1977).

[8]The CRBR was finally killed when the nuclear industry refused to contribute more to the escalating construction cost and the congressional legislators recognized that the breeder was unlikely to be economically competitive with current nuclear power reactors until well into the 21st century. See, for example, Chow (1983).

has about 100 tonnes of capacity.[9]   Other countries, including the United States and Germany, have no reprocessing plants operating currently.

As sensitive plutonium operations in the civilian fuel cycles world-wide face restructuring, the plutonium from dismantled nuclear weapons is becoming a new factor to consider.   Different parties evaluate the situation differently.   For example, the Japanese and German governments see profit opportunities in helping FSRs get rid of weapon-grade plutonium by burning it in thermal or fast reactors.[10] On the other hand, FSRs think that plutonium is valuable and that they can sell it.   This difference in perceived value creates an obstacle in agreeing on what to do with weapon-grade plutonium. Allowing the burning of weapon-grade plutonium might help the plutonium-use advocates keep some facilities, such as MOX fuel fabrication plants, operating or under construction until the day of massive plutonium commercialization.   Thus, weapon-grade plu-tonium burning and reactor-grade plutonium use must be distinguished under certain important situations.   For example, burning weapon-grade plutonium in FSRs should be to quickly and easily eliminate the kind of plutonium that can be refashioned into nuclear weapons.   In no way should that activity be considered as a support of plutonium use in general.

AN ELECTRICITY COST MODEL

In the post-Cold War era, economics has played an increasingly im-portant role in shaping nations' foreign policies.   If using reactor-grade plutonium (plutonium from civilian spent fuel) in current nu-clear reactors will generate enormous economic benefits, one cannot realistically expect countries to forgo it for proliferation concerns. Further, if reactor-grade plutonium is used widely, the added risk in using weapon-grade plutonium will be relatively small.   On the other hand, if thermal recycle and fast reactors are uneconomical, a coun-try would be more reluctant to plunge into a plutonium economy.

---

[9]*World Inventory,* pp. 90 and 106.

[10]In this report, plutonium burning in reactors has the same meaning as using plutonium as fuel in reactors.

We have developed a model that uses discounted cash flows to determine the cost of generating a given amount of electricity from a specific type of nuclear reactor or fuel cycle. For example, we determine the busbar cost in mills per kilowatt-hour of electricity from a nuclear pressurized water reactor (PWR) with no plutonium recycle (a once-through system), which is the most widely used type and mode of commercial nuclear power plant worldwide.[11] We also determine the cost of generating electricity by thermal recycle. The electricity cost depends partly on the uranium price. The model can be used to determine the breakeven uranium price at which electricity from thermal recycle costs the same as that from the once-through system. The higher the breakeven price, the more uranium will be available at or below that price, and it will take longer for thermal recycle to become economical. From our knowledge of the amount of available uranium resources, we will estimate the uranium depletion rate and the year that the uranium price is expected to reach the breakeven price. This will help us estimate the year that thermal recycle will first become economical. We will use the same approach to estimate when the fast reactor will become economical.

The key parameters used in our calculations are shown in Table 3.1.[12] To better reflect the tradeoff between uranium enrichment and raw materials, we used in the calculation a tail[13] optimized for the lowest combined cost. In other words, as the uranium price increases, countries will discard the uranium during the enrichment process at a lower fissile isotopic concentration to conserve yellowcake. Since the cost of plutonium-containing fuel fabrication and that of plutonium reprocessing are two key determinants of plutonium economics and will have an important influence on the

---

[11]The busbar cost is directly related to fuel, operating and maintenance, and power plant capital expenses in electricity generation, excluding costs such as marketing and distributing to customers. In this report, the electricity cost is the same as the busbar cost. One mill equals one-thousandth of a dollar.

[12]Unless noted otherwise, all costs in this report are in 1992 U.S. dollars.

[13]During uranium enrichment, the isotopic concentration of U-235 will increase in some uranium and decrease in some. The uranium with the lower level of U-235, which is called the tails, will be taken out of the enrichment process. The level of U-235 in the tails is called the tails assay.

Table 3.1

**Key Parameters for Calculating Plutonium Value
(in 1992 U.S. dollars)**

	LWR (OT)	LWR (PM)	LWR (FM)	LMFR
Plant capital cost, $/kWe	1,900	1,900	1,900	2,020
Plant capacity factor	0.7	0.7	0.7	0.7
Enrichment, $/SWU	70	70	—	—
UO_2 fabrication cost, $/kg HM	200	200	—	—
MOX or fast-reactor fuel fabrication cost, $/kg HM	—	800	800	960
Spent fuel/waste transportation cost, $/kg HM	30	20	20	50
Reprocessing cost, $/kg HM	—	450–1,600	450–1,600	540–1,920
Spent fuel/waste disposal cost, $/kg HM	550	550	550	550
Separated plutonium storage cost,[a] $/kg-yr	—	430	430	430
Annual requirement, kg Puf/GWe-yr	—	320	1000	960
Annual plutonium discharge, kg Puf/GWe-yr	166	320	646	1060
Annual charge, MTHM/GWe	26.8	26.8	26.8	34.4
Annual discount rate, %	10	10	10	10
Plant life , years	30	30	30	30

NOTE: LWR (OT) = light water reactor (once through), LWR (PM) = LWR (with one-third or partial MOX fuel), LWR (FM) = LWR (with 100 percent or full MOX fuel), LMFR = liquid metal fast reactor, SWU = separative work unit, Puf = plutonium fissile, HM = heavy metal, i.e., uranium or plutonium, MTHM = metric ton of HM, GWe = gigawatt-electric.

[a]Marginal storage cost for FSR weapon-grade plutonium in the United States.

recommended policy in plutonium management, we will discuss these two costs separately below.

MOX FABRICATION CAPACITY AND COST

In this section, we examine two questions. If plutonium from dismantled nuclear weapons is to be used in nuclear reactors, will there be enough MOX fabrication capacity to incorporate all of the plutonium in MOX? Why is MOX fabrication cost high and why will it remain high relative to UO_2 fabrication cost? Both questions have a direct bearing on the policy about what to do with weapon-grade plutonium.

Figure 3.1 shows current and future MOX fabrication capacity world-wide. Annual capacity is 15,000 kg of Puf in 1993 and is expected to increase to 17,400 kg by 1995 and to 34,500 kg by 2000. Various countries have planned on this capacity for their use of reactor-grade plutonium in thermal and fast reactors during the next 10 years and beyond, without anticipating the added military plutonium from FSR and U.S. dismantled bombs. Over the next 10 years, this military plutonium will, however, amount to 23,000 kg Puf per year. Even FSR materials alone will amount to 13,000 kg Puf per year.[14]

If countries follow their current plan of using reactor-grade pluto-nium in thermal reactors and fast reactors, and if, in addition, coun-

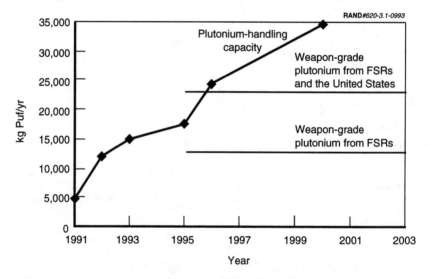

SOURCE: Plutonium-handling capacity is based on data in *World Inventory*, p. 121.

Figure 3.1—Plutonium Fuel Fabrication Capacity Worldwide

---

[14]We assumed for this discussion that weapon-grade plutonium will be used in thermal reactors and in equal annual amounts from 1995 to 2005. Our analysis and our conclusions will be similar if part of the weapon-grade plutonium is used in fast reactors.

tries want to burn weapon-grade plutonium, additional MOX fabrication capacity will be needed. An alternative is to replace reactor-grade plutonium with weapon-grade plutonium and reduce correspondingly the amount of plutonium to be separated from spent fuel. The advantage of this scheme is that it helps to eliminate weapon-grade plutonium; the disadvantage is the placing of weapon-grade plutonium in nonnuclear weapon states, such as Germany and Japan, if they participate in the plutonium burning program. Burning weapon-grade plutonium in FSRs will incur the advantage but not the disadvantage. Unfortunately, as shown in Table 3.2, FSRs have planned a capacity of merely 5 tonnes of MOX a year or about 300 kg Puf a year sometime in the 1990s. That capacity is far from adequate to handle the expected amount of surplus weapon-grade plutonium (about 13,000 kg Puf a year) in FSRs. Either more plants have to be built or the fabrication and even the burning will have to take place outside FSRs.

An alternative is to burn weapon-grade plutonium, instead of reactor-grade plutonium, in France and the United Kingdom, the two nuclear weapon states with the most extensive infrastructure for handling plutonium. France can currently handle about 2,200 kg of Puf/yr and the United Kingdom 400 kg of Puf/yr. Current MOX fabrication capacity is insufficient to handle the 13,000 kg of Puf a year from FSR surplus weapon materials. On the other hand, France expects an additional capacity of 6,900 kg Puf/yr by 1996 and the United Kingdom another 4,200 kg in the late 1990s. Thus, by the latter part of the 1990s, France and the United Kingdom combined should be able to use all their MOX fabrication capacity, which is intended for reactor-grade plutonium, for FSR weapon-grade plutonium. We will return to this option in Chapter Four.

Putting aside the possibility of using weapon-grade plutonium in reactors, it is important to know whether thermal recycle and breeders will be economical soon. If they will, and if they are massively deployed, the proliferation risk will be so high that using weapon-grade plutonium to support them does not make a bad situation much worse. We will show, however, that recycle and breeders will not be economical for a long time. But first, we want to examine the current and future costs of MOX fabrication, which is a key determinant of whether plutonium use will be economical soon.

Table 3.2

Worldwide Capacity of Plutonium Fuel Fabrication

Status/ Country	Facility	Period of Operation	Fuel	Capacity (tonnes of MOX/year)	Maximum plutonium consumption (kg Puf/year)
Operating					
Belgium	Diesel DEMOX	1973–	FBR/LWR	35	1,600
France	Cadarache ATPu	1970–1989	FBR	15	4,000
France	Cadarache CFCa	1990–	FBR / LWR	10 / 15	2,200
FRG	Hanau BEW 1	1972–1992?	FBR/LWR	25–30	1,050
Japan	Tokai PFFF	1972–	FBR / ATR	1 / 9	300
Japan	Tokai PFPF	1988–	FBR	5	900
United Kingdom	Sellafield	1970–1989	FBR	4	1,300
Planned					
Belgium	DEMOX P1	Mid-1990s?	LWR	35	2,100
France	Marcoule Melox	1996–	LWR	115	6,900
FRG	Hanau BEW 2	1992–	LWR	80–120	7,200
Japan	Tokai PFPF	1993/1994–	ATR	40	2,400
Japan	Rokkasho	Late 1990s?	LWR	100?	6,000
Russia	Chelyabinsk	1990s?	FBR	5?	300?
United Kingdom	Sellafield MDF	1993–	LWR	8	400
United Kingdom	Sellafield SMP	Late 1990s?	LWR	50–70	4,200

SOURCE: Albright et al. (1993), p. 121.

Inherent physical differences make MOX fuel more expensive to fabricate than UO_2 fuel. Pu-239 is much more radioactive and more thermally hot than U-235.[15] This higher radioactivity and thermal heat in MOX fuel are also due to the presence of Am-241, the radioactive decay products of Pu-236 and U-232, the heat output from

---

[15]For example, the alpha activity of U-235 is 0.0000021 curie/gram whereas that of Pu-239 is 0.0613 curie/gram (see Benedict et al. (1981), p. 429), 30,000 times more intense.

Pu-238, and the neutron activity from spontaneous fission and alpha-n reactions.[16] Plutonium as an alpha emitter also offers a very serious health hazard. The major danger arises from the fact that plutonium entering the body via the respiratory system initially locates in the lungs. Although larger particles of plutonium tend to leave the lung within a few days in mucus and are almost completely eliminated through the intestines, smaller particles, especially those reaching the deeper parts of the lung, stay over a period of years. Some plutonium is then dissolved in the blood and is deposited in the bone marrow and the liver. The biological half-life of plutonium in the bone marrow is about 100 years and in the liver about 40 years, where even very small amounts can cause cancer.[17] Isolation from respiratory intake is accordingly more stringent for areas where plutonium is present than for areas containing only uranium.[18] MOX fabrication plant layouts and physical controls are designed to ensure that present radiation and health protection standards are met for both workers and members of the general population. MOX handling requires substantially greater use of air locks and isolation facilities. In a MOX fabrication facility, working areas may from time to time be contaminated with plutonium. This requires that equipment be dismantled and cleaned more regularly. This cleaning process requires isolation and may require part of the plant to be closed. Workers must access the isolated area through air locks in fully protective clothing with an external air supply. In contrast, a UO_2 fabrication facility used for current reactor fuels does not involve plutonium and requires only less-expensive, conventional protective equipment to handle accidental contamination. Moreover, MOX fabrication involves total quantities of plutonium many times above critical mass.[19] Fuel must be handled in small enough batches (less than about 2 kg of plutonium) and with care to avoid criticality problems. These precautions and the added need to protect plutonium inventories from diversion inevitably make MOX handling much more expensive than UO_2 handling.

---

[16]Benedict et al. (1981), pp. 364, 366–370, 372–376, and 448.

[17]Partington (1993), pp. 127–155.

[18]See 10 CFR (Code of Federal Regulations), Chapter 1, Part 71, Table A.1, July 1992.

[19]The MOX fuel for an annual reload for a LWR (PM) contains about 55 times the critical mass if weapon-grade plutonium is used, or 66 critical masses if reactor-grade plutonium is used.

Some West European estimates for MOX fabrication have ranged from \$1,300 to \$1,600/kg HM.[20] A contract was reportedly concluded between Japan and Belgium for fabrication of 500 tonnes of MOX fuel at a price of about \$1,400/kg of MOX.[21] Other estimates claim contracted fabrication costs as high as \$3,000/kg of MOX, depending on how fully the capacity of the MOX fabrication is used.[22] So there is a substantial degree of uncertainty in the MOX fabrication cost. By comparison, UO_2 fabrication costs have been estimated at only about \$200 to \$250/kg.[23] Thus, the difference in MOX and UO_2 fabrication costs currently runs from \$1,050 to figures much higher.

An OECD study group estimated that, as MOX fuel is used on a more significant scale, MOX fabrication costs will decline. The study predicted a 15 percent to 25 percent decrease in MOX fabrication cost from the early 1990s through the late 1990s and into the early part of the 21st century. But such a decrease will still leave MOX cost significantly higher than that of UO_2. To be conservative, we used in this study a difference of only \$600/kg HM.[24] Our figures for MOX and UO_2 fabrication costs and their difference are the same as those used by the Nuclear Energy Agency Expert Group.[25]

Electricity from thermal recycle would have been much cheaper if only the much higher fabrication cost for the 0.32 tonne of reactor-grade plutonium involved in the annual reload of a reactor (LWR (PM)) had to be paid. In reality, this 0.32 tonne of plutonium is mixed with a large amount of uranium dioxides to yield 8.2 tonnes of

---

[20]Berkhout and Walker (1990), p. 22. MOX fabrication costs have been given in \$/kg HM and \$/kg MOX. Since oxygen is much lighter than heavy metals, the difference between the two units is not large. We use both terms and follow whichever term is used in the quoted reports.

[21]Vermeulen (1993).

[22]Leventhal and Dolley (1993), pp. 24–25.

[23]Private communication with Michael Schwartz, Energy Resources International, Washington, D.C. , on June 22, 1993.

[24]For thermal reactors, we used \$800/kg HM for MOX fabrication and \$200/kg HM for UO_2. For fast reactors, we used a MOX fabrication cost that is 20 percent higher, because of higher plutonium concentration in fuels for the fast reactor cores.

[25]NEA gave an "illustrative cost range" of \$700–\$1,000/kg HM for MOX fabrication with their computation based on \$800/kg HM. The corresponding range for UO_2 fabrication is \$175–\$300/kg U with computation based on \$200/kg U (NEA, 1989, p. 58).

MOX fuel. In other words, the higher fabrication price is paid for the whole 8.2 tonnes, instead of for merely 0.32 tonne.

REPROCESSING COST

In this section, we examine the reprocessing prices being charged by the United Kingdom and France—the two countries that have made the most substantial financial and other commitments to reprocessing. Moreover, because most of the reprocessing plants in the United Kingdom and France have been built and the capital costs have already been sunk, we will estimate the reprocessing cost including only the operating and maintenance cost.

The argument about sunk costs has been used often in the past to keep uneconomical projects alive. Project advocates underestimate the ultimate total cost of a project, simply because they do not include sunk costs in the analysis of whether a project should be terminated. For this reason, we should be very cautious about projected decreases in future costs, when a project's history shows rampant cost overruns. Despite repeated projection of declining reprocessing costs, the record shows just the opposite.

Both reprocessing costs with and without sunk costs are relevant to this study, and in this section we will estimate both of them. The reprocessing cost without the sunk cost is the pertinent parameter for deciding whether countries should terminate or scale back reprocessing during the operating life of their existing plants. On the other hand, the capital cost (from which the sunk cost is determined) should be included in determining when reprocessing will become economical. As we shall show, reprocessing will not be competitive for at least 30 years; existing reprocessing plants will have been retired by then, and new plants will have to be built. Thus, the capital costs of new plants are relevant to the date of economic reprocessing.

The total costs for reprocessing spent fuel have been examined in a number of studies. NEA estimated that, for large plants handling 1,200 tonnes per year,[26] a price of about $750/kg HM (excluding the

---

[26]Each GWe LWR generates about 27 tonnes (HM) of spent fuel a year.

cost of short-term storage of spent fuel) would provide a 10 percent return on capital.[27] The price would have to be $1,000/kg HM if the plant were smaller or to account more conservatively for the technical uncertainties. An April 1991 study reported that the French La Hague and the British Sellafield have been charging their customers about $1,400 to $1,800/kg HM. But it also reported that a price of $900/kg HM has been offered for reprocessing in the post-2000 period. This reduction probably comes from the belief that much of the capital expenses will have been recovered by then. In other words, the reduction is not due to a lowering of the operating and maintenance costs. For the present study, a $900/kg HM is used as the total reprocessing cost, although the current charge is much higher.

Further, we want to develop an even lower cost to represent the situation where the capital cost has been sunk. This calls for determining the portion of reprocessing cost that is attributable to capital cost and the portion to operating and maintenance. The experience at the U.K. Thorp reprocessing plant gives us one such estimate. The capital cost of the Thorp plant has been reported to be $2.8 billion (at an exchange rate of £1 = $1.5). With a 25-year expected plant life and 700 tonnes of annual throughput,[28] the capital cost component amounts to $440/kg HM. Using the post-2000 figure of $900/kg HM for reprocessing cost, we estimated that the capital cost amounts to about half of the reprocessing cost, and the operation and maintenance costs account for the remaining half or $450/kg HM. We use this figure to represent the reprocessing cost when the capital cost is sunk.[29]

Another question of interest is the cost of newer reprocessing plants. If the cost is lower, reprocessing will be more economical in the

---

[27]NEA (1989), p. 62.

[28]Berkhout and Walker (1990), pp. 8, 10, 30, and 54.

[29]These operation and maintenance costs may be lower than actual and thus are conservative for the purpose of this study. If we had used the current charge of $1,400 to $1,800/kg HM (excluding charges for long-term high-level waste storage, transport, and disposal) as the total reprocessing cost, the capital cost of $440/kg HM would have accounted for substantially less than half of the total. In other words, even ignoring capital cost, the reprocessing cost would be considerably more than the $450/kg HM figure that we used.

future. Recent data indicate otherwise, however. Japan Nuclear Fuel Limited (JNFL)'s 800 tonnes a year commercial reprocessing plant at Rokkasho had a ground breaking on April 28, 1993, and is expected to be in operation around the year 2000. The plant's capital cost is expected to be $7.5 billion.[30] At a 10 percent rate of return and 30 years of plant life, this capital cost translates into $1,000/kg HM, more than twice that of Thorp. Therefore, the capital cost for future plants is not expected to be less at least for the next decade. JNFL did not release the operation and maintenance costs.

The two reprocessing costs used in our analysis are $450/kg HM and $900/kg HM, which are very similar to the "illustrative cost range" of $500 and $1,000/kg HM used by the NEA.[31] The $450/kg HM figure is used as the reference cost for using existing reprocessing plants, whose plant capital costs have been paid or sunk. The $900/kg HM figure is used as the reference total reprocessing cost for determining when thermal recycle and fast reactors will become economical. Moreover, since we are looking 30 years and more into the future, and we cannot ignore the possibility that capital and operation and maintenance costs might fall eventually, we use $450/kg HM also as a lower bound to the total reprocessing cost for those long-term calculations.

ECONOMICS OF CIVILIAN PLUTONIUM USE

In this section, we discuss when the use of plutonium recovered from spent fuel will be economical. The economics of military plutonium in civilian reactors will be discussed in Chapter Four.

Figure 3.2 shows the breakeven price of thermal recycle under current and future reprocessing costs. This is the yellowcake price at which electricity generation from thermal recycle is as costly as that from current once-through reactors. We calculate that at the current

---

[30]"Construction Begun for Plant in Japan" (1993), p. 82, and "French Reprocessing" (1993), p. 14.
[31]NEA (1989), p. 58.

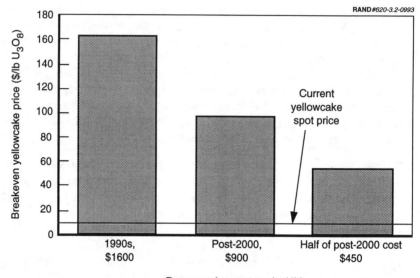

Figure 3.2—Thermal Recycle Breakeven Price

reprocessing charges of around \$1,600/kg HM[32] by France and the United Kingdom, the breakeven price would be \$160/lb U_3O_8 of yellowcake, or 16 times the current spot price of merely \$10/lb.[33] Even with reprocessing services costing \$900/kg HM after the year 2000, the breakeven price is still a high \$95/lb. Finally, as the lower estimate, we assume that the capital cost of a reprocessing plant has

---

[32]The midpoint of the range of \$1,400 to \$1,800/kg HM that Japanese and other customers pay.

[33]In this report, we use metric units, except for the price and amount of natural uranium. The price and amount are often quoted in British units in the trade journals, especially in the United States. One tonne (or MT) equals 1.1 ton (or short ton). The spot price fluctuates and was recently as low as \$7.5/lb. For this report, we use a more conservative number of \$10/lb. We use the spot price instead of the average price for existing long-term contracts, because the spot price better reflects the latest estimate of the future price. Since the average long-term price includes contracts signed long ago, that price does not represent the current uranium market situation and expectation. Another good indicator is the average price for long-term contracts recently signed. Unfortunately, this average price is not generally available. Around the end of 1990, the price on long-term contracts was \$12/lb of yellowcake (Berkhout and Walker, 1990, p. 22).

been sunk or that the total reprocessing cost amounts to half of the post-year-2000 cost, or only \$450/kg HM. The yellowcake price will still have to rise to \$50/lb before reprocessing is economical. The breakeven price is five times the current spot price. We will estimate when uranium prices will reach various breakeven prices after we discuss breakeven prices for fast reactors.

Note that the \$450/kg HM figure and the derived \$50/lb figure, which we rely on heavily for later discussion and for making policy recommendations, are very conservative for two reasons. First, \$450/kg HM is the lowest of the three reprocessing costs and is the most optimistic case for plutonium use. In fact, as discussed above, the reprocessing industry's offer to lower the price from \$1,400–\$1,800/kg HM to \$900/kg HM post-2000 might have already assumed that the plant capital cost will have been recovered by the end of the century. In that case, it is unlikely that the reprocessing cost will further be lowered to \$450/kg HM. Second, the experience world-wide in reprocessing cost in particular and in nuclear electricity gen-eration in general has been rising because of environmental, regula-tory, and operational concerns. Prediction of declining cost has consistently been proven wrong.

Figure 3.3 shows the breakeven yellowcake price for a fast reactor as a function of the plant capital cost ratio of fast reactor to LWR. Because it is more difficult to handle opaque and volatile liquid sodium and large quantities of plutonium, there is general agreement that the power plant capital cost of a fast reactor is inherently higher than that of a LWR. In the past, the power plant capital cost ratio of fast reactor to LWR has been estimated to be 20 percent to 100 percent higher.[34] Here, we assume that the power plant capital cost of a fast reactor is only 10 percent to 20 percent higher. At 10 percent, the breakeven price for a fast reactor to be competitive with a LWR is from \$140 to \$170/lb U_3O_8 of yellowcake. The lower value assumes that the reprocessing cost is \$540/kg HM.[35]

---

[34]Chow (1979), p. 47.

[35]We assume that the reprocessing cost for fast reactor fuel is 20 percent higher than that of LWR spent fuel, because of higher plutonium concentration in the spent fuel. DoD's Nonproliferation Alternative Systems Assessment Program (NASAP) estimated

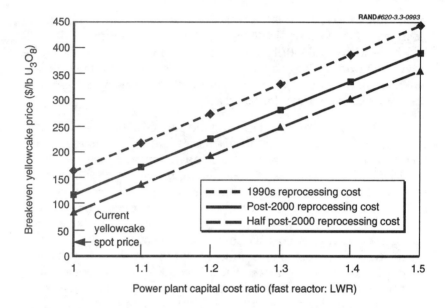

Figure 3.3—Fast Reactor Breakeven Price

In other words, even if one argues that the capital cost of the reprocessing plant has been paid for, just the operation and maintenance reprocessing costs will lead to a breakeven price of $140/lb of yellowcake. To obtain the higher breakeven price, we used a reprocessing cost of $1,080/kg HM,[36] which assumes that the cost will come down substantially from the current level but still includes reprocessing plant capital cost. With a fast reactor at 20 percent higher power plant capital cost, the breakeven price ranges from $190 to $220/lb of yellowcake.

When will the yellowcake reach various breakeven prices? Factors that affect uranium price include the amount of uranium resources available, their extraction costs, and their nuclear power growth.

---

the fast reactor reprocessing cost to be 20 percent to 30 percent higher (Chow, 1979, pp. 31–32).

[36]This is 20 percent higher than the LWR reprocessing cost of $900/kg HM discussed in the thermal recycle case.

Current nuclear capacity worldwide is 343 GWe.[37] As for future nuclear capacity, we use the projection prepared by the NEA and the IAEA.[38] NEA's projection is generally considered to be one of the most authoritative in the field. Its latest projection to the year 2010 is based on plants under construction and planned. Since it takes a decade or more to plan and build a nuclear power plant, it is unlikely that actual nuclear capacity by the year 2010 would be higher than its projection.[39] NEA projected the worldwide nuclear capacity to be 415 GWe by the year 2000 and 474 GWe by 2010. It should be noted that, as with many other of their projections in the past, NEA tends to overestimate.[40] Although we do not have enough past worldwide projections to prove the overestimation, OECD data, which are more available, show the same trend. In 1977, NEA projected that the nuclear capacity in OECD countries by 2000 would be 1,200 GWe. As the year 2000 draws near, the projection declines precipitately. By 1982, the projected year-2000 capacity had dropped to 500 GWe and by 1987, to 340 GWe. The current projected OECD nuclear capacity for the year 2000 is 300 GWe.[41] The year-2000 projection was reduced by a factor of four in sixteen years.

NEA has not projected beyond the year 2010. Projecting nuclear capacity more than 20 years from now is highly uncertain, but one must look far beyond 20 years when deciding on the policy of future nuclear development. We have used two projections. In the reference nuclear growth case, we assume that worldwide nuclear capacity beyond 2010 will grow at the same rate as the projected growth during the period 1992–2010—1.8 percent a year (Figure 3.4).

---

[37]Current capacity as of August 31, 1992. Nuclear Engineering International (NEI) (1993), p. 12.

[38]NEA and IAEA (1992), pp. 58–59; and NEA (1992), pp. 38 and 42.

[39]The actual construction time varies by country. In the United States, it takes on average 12 years; in France, 7 years; and in Japan, 5 years (NEI, 1993, p. 12). If planning time is added, the total time in many countries is likely to be about a decade or more.

[40]In fact, another authoritative source, NEI, estimated that world capacity by the year 2000 will be only 400 GWe or about 15 GWe lower than the estimation by NEA (NEI, 1993, p. 14).

[41]NEA (1992), p. 42. Thus, nuclear capacity outside OECD by the year 2000 should be 115 GWe to match the projected total worldwide nuclear capacity of 415 GWe by 2000.

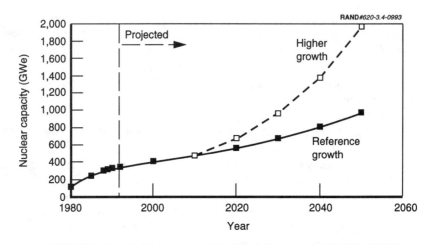

SOURCES: Historical data are from *The World Almanac*, p. 187; IAEA (1992), p. 56; and NEI (1993), p. 10. Projection to 2010 was from NEA and IAEA (1992), pp. 58–59, and NEA (1992), pp. 38 and 42. Projections after 2010 were by the authors.

Figure 3.4—Projected Nuclear Capacity Growth Worldwide

In the higher nuclear growth case, we assume that the growth rate will be twice the reference rate—3.6 percent a year. In contrast, the (nuclear and nonnuclear) electricity capacity growth in OECD during 1991–2010 is projected to be merely 1.5 percent.[42] Thus, we have assumed that nuclear capacity will grow at a rate that is more than twice the total electric capacity growth. Nuclear capacity may grow even faster if the greenhouse effect and air pollution become serious problems. If there is a drastic increase in nuclear capacity, however, there will be plenty of advance warning, because many countries generally take a decade or more to plan and build nuclear power plants.

The number of plants (assuming 1 GWe per plant) to be built, however, will account for more than the capacity growth, because additional plants will have to be built to replace those retired during the period. Currently, a 1 GWe LWR consumes about 5,500 tons of yellowcake over its 30-year lifetime. We assume that, on average, plants

---

[42]NEA (1992), p. 38.

coming into service after the year 2000 will use 20 percent less uranium through evolutionary technological improvements. We make another assumption that, as soon as a plant begins operation, the whole lifetime uranium requirement will be secured through a long-term contract and that the requirement will be subtracted from the uranium resources immediately. This is a very conservative assumption to assure that the uranium requirement will be met.

On the uranium supply side, we have based our estimates on the yellowcake included in the Reasonably Assured Resources, the Estimated Additional Resources I and II, and the Speculative Resources categories estimated by NEA and IAEA.[43] NEA/IAEA grouped yellowcake in the Speculative Resources under three cost ranges: up to $50/lb U_3O_8, between $50 and $100/lb, and unassigned cost range but below $100/lb. To be somewhat more conservative, we group all Speculative Resources into the cost range of $50 to $100/lb of yellowcake. We assume that the lower-cost uranium resources will be recovered first.

Finally, we convert the above extraction cost to the yellowcake price charged to the utilities. NEA/IAEA cost estimates do not allow for return on investment,[44] as they should. These costs are estimated the same way as the forward cost used by the U.S. Department of Energy in its estimates of U.S. uranium resources. John Klemenic, chief of the Supply Evaluation Branch of the Grand Junction Office of the Energy Research and Development Administration (predecessor of the Department of Energy) performed extensive analysis on the conversion of forward cost to "full economic cost." The latter

---

[43]The amount of uranium resources is generally given in three cost ranges: up to $30/lb U_3O_8, $30–$50/lb, and $50–$100/lb. Reasonably Assured Resources (RAR) include uranium in known mineral deposits. Estimated Additional Resources (EAR) I include uranium, in addition to RAR, inferred to exist in extensions of well-explored deposits. EAR II is uranium, in addition to EAR I, that is expected to occur in well-defined geological trends or areas of mineralization with known deposits. Speculative Resources include uranium, in addition to EAR II, that is thought to exist in deposits discoverable with existing exploration techniques. See NEA and IAEA (1992), pp. 13–29, for a detailed definition and estimates of the amount of uranium in these categories. We have also included uranium in the so-called "Other Known Resources" category, which includes resources in Chile, China, India, Romania, and the FSRs. Although there are 1,080,000 tonnes of such resources, we include only 890,000 tonnes, because we assume that 20 percent of FSR resources have been exhausted.

[44]NEA and IAEA (1992), p. 14.

included a rate of return typical of that of the mining companies[45] and is a better proxy for price. To provide a rate of return of 15 percent, the full economic cost is 30 percent to 70 percent higher than the forward cost, depending on the ore grade (or percentage of U_3O_8 in the ore). After further analysis, Vince Taylor used a factor of 1.7 to convert the cost of high-grade yellowcake to price and a factor of 1.5 to convert the low-grade yellowcake.[46] To be conservative, we used a factor of 2 for all cost-to-price conversion.

Figure 3.5 shows the cumulative amount of uranium available up to a given price. For example, there will be 8 million tons of U_3O_8 available worldwide at or below \$100/lb. Considering that the lifetime requirement of each future GWe plant, LWR (OT), is 4,400 tons, we found that the resources at \$100/lb can fuel 1,800 such plants, or five times the current nuclear capacity. In the next two sections, we will first discuss extracting uranium from seawater and then estimate when uranium prices will reach various breakeven prices at which thermal recycle and fast breeders will become economical.

EXTRACTING URANIUM FROM SEAWATER AS A POSSIBLE BACKSTOP TECHNOLOGY

Uranium is present in seawater at 3.34 milligrams per cubic meter.[47] The vast amount of water in the world's oceans could yield 4 billion tonnes of uranium, 1 percent of which would provide enough fuel to meet the uranium requirement at the current level for 700 years. Although nuclear capacity may grow over time, even 1 percent could last for many years. For example, in our high nuclear growth case (Figure 3.4), which assumes exponential nuclear growth at 3.6 percent indefinitely, the 1 percent *alone* could fuel the lifetime requirements of all plants built in the next 95 years. In other words, we could continue to use the current type of nuclear powers in the once-through mode with merely 20 percent uranium efficiency improvement until the year 2088. Nuclear capacity by then would be 7,600

---

[45]Klemenic (1972, 1974). Both reports were quoted in Taylor (1975).

[46] Taylor (1975), Appendix B.

[47]Our discussion in this section is based on Benedict et al. (1981), pp. 261–264.

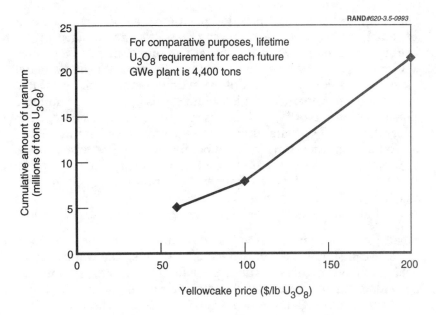

Figure 3.5—Uranium Resources Worldwide

GWe, about 22 times the current level! Thus, seawater has the potential to supply uranium for almost one hundred more years.

One tonne of uranium, however, requires the processing of 300 million cubic meters of seawater. This is the key problem. According to a 1974 Battelle Pacific Northwest Laboratory report and as summarized in Benedict et al., the most promising process for extracting uranium from seawater is the selective adsorption of uranium on hydrous titanium oxide (titania). After their review of the literature, they concluded: "It seems likely that extraction of uranium from sea water would cost on the order of $500/lb U_3O_8." Since their analysis was performed in 1981, expressing the cost in 1992 dollars results in $750/lb U_3O_8.

On a purely economical basis, plutonium use will probably be economical well before the yellowcake price reaches $750/lb U_3O_8. Moreover, a lot more uranium than we have assumed for this study will be found in the ground, and there is no need to use uranium from seawater.

On the other hand, current knowledge of extraction technologies is extremely limited, because uranium prospectors do not have any incentives to explore when the current uranium price is less than $10/lb U_3O_8. We believe, however, that governments should better assess the feasibility and cost of extracting uranium from seawater. With the breakeven price for plutonium use as high as a few hundred dollars per pound of U_3O_8, it is possible that better technology can lower the cost of seawater uranium to the same level. If the price of seawater uranium turns out to be below the breakeven price, seawater uranium can serve as a backstop that could be used to fuel the current type of once-through proliferation-resistant reactors indefinitely. The world might not have to use proliferation-prone plutonium in nuclear fuel cycles for the next hundred years or more.

Finally, even if seawater uranium costs more than the breakeven price, it can still be used in two situations. First, the world may be willing to pay a higher electricity price for a lower proliferation risk. Second, it can be used during the transition period to massive use of plutonium. Although uranium is expensive, using high-cost uranium might still be cheaper than closing down the plants or using other nonnuclear electricity-generating plants.

WHEN WILL THERMAL RECYCLE AND BREEDERS BE ECONOMICAL?

The timing of economical plutonium use depends on how fast the uranium price will rise. Before we project the price growth, we examine the historical trend in uranium prices for clues to a future pattern (Figure 3.6). In the 1950s and then again in the 1970s, there were peaks. In 1992 dollars, the former peak reached $70/lb U_3O_8 and the latter peak, $95/lb U_3O_8.

The earlier price peak could be explained by the fact that in 1948 the U.S. government began a major uranium-buying program to support the U.S. nuclear weapon buildup while the uranium industry was still in its infancy.[48]   This program drove up the yellowcake price,

---

[48]This discussion on the causes of the two peaks is extracted from Taylor (1975), pp. 76–85.

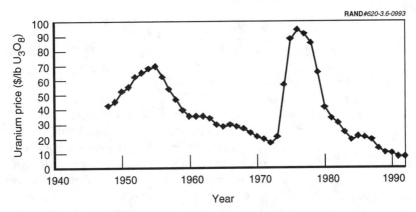

SOURCES: Heritage Foundation (n.d.). The curve there in 1982 dollars was based on 1948–1967 data (average U.S. contract prices) from the U.S. Department of Energy and 1968–1975 data (spot prices) from NUEXCO (1982), p. 28. The data for 1976–1990 (NUEXCO exchange values) were from Combs (1991), Figure 6. The data for 1991–1992 (NUEXCO exchange values) were from *Nuclear News* (1992), p. 55, and *Nuclear News* (1993), p. 82. The authors converted all data to 1992 dollars.

Figure 3.6—Historical Trend of Uranium Prices (in 1992 dollars)

because the supply could not meet the sudden surge in demand. The decline in uranium prices after 1955, however, showed that the uranium industry could respond quickly to higher prices and higher incentives. The increased exploration and improved production technology helped to drive the price down.

The cause of the second peak was different. Before 1973, the U.S. Atomic Energy Commission (AEC) permitted customers to obtain enrichment services without requiring the needed quantities of yellowcake to be delivered by specific dates. Customers needed only to give AEC 180 days advanced notification of enrichment needs. In 1973, AEC announced that henceforth all contracts would require that customers buy specified quantities of enriched uranium at specified dates and deliver the natural uranium required for the enrichment services. Worse yet, AEC also created an overdemand of enrichment services, and therefore of uranium, by announcing that, until June 1974, it would issue contracts only to customers who needed enrichment for initial cores before June 1982. Since AEC was at the time the only assured source of enrichment, customers rushed to sign the new contracts, even if they might not need enriched uranium before June 1982. Then the oil embargo in 1973 caused utilities to want long-term assurance of fuel, including uranium. The forward purchase of uranium rose from 6,000 tons per year during 1968–1972 to 34,000 tons in 1973.[49] Finally, the U.S. government in January 1973 proposed increasing the U-235 isotopic concentration in the waste stream (tails assay) from 0.20 percent to 0.275 percent in 1976 and to 0.30 percent in 1981. A change from 0.20 percent to 0.30 percent would force customers to deliver 28 percent more uranium and thus raised uranium demand. Then, by 1975, plans for early reprocessing and recycle of plutonium also appeared doubtful. The Energy Research and Development Administration (the successor of AEC) compounded the uranium shortage problem by announcing that it would raise the tails to 0.37 percent if recycle were not approved. That would raise the customers' uranium delivery by 37 percent instead of 28 percent. The interplay of these factors pushed the price of the uranium to as high as $95/lb U_3O_8 by 1976.

The cause of the first peak—the beginning of the nuclear weapon buildup program—will not repeat in the future. The cause of the second peak was a series of flawed government policies on top of an energy crisis and a unique situation. Flawed policies and energy crises will recur in the future, but the United States will never again be the only enrichment service provider; there are now commercial enrichment facilities in France, the FSRs, the Netherlands, the United Kingdom, and other countries. Peaks in uranium price will appear in the future and might even be quite pronounced, but they are unlikely to be permanent. The world's long-term major decision about civilian nuclear power, such as plutonium use, should be driven in part by the general trend of uranium prices, not by their temporary peaks and valleys. In the rest of this section, we use our projected uranium supply and demand to determine the price trend.

In the reference nuclear growth case (1.8 percent per year), we estimated that thermal recycle is not expected to be economical until sometime between 2025 and 2048 (Figure 3.7). The key factors driving the poor economics of thermal recycle are the high

---

[49]Lorie and Gody (1975), p. 39.

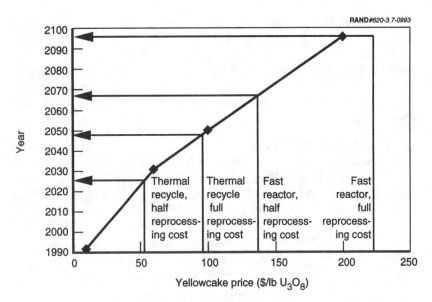

Figure 3.7—When Plutonium Thermal Recycle and Fast Reactor Become Economical (Reference Nuclear Growth Case)

fabrication cost of plutonium-bearing MOX fuel and the reprocessing cost. We explained above that the MOX cost is much higher than the cost for fabricating the typical uranium-dioxide fuel for current reactors, because of the criticality, toxicity, and high radioactivity of plutonium. The high reprocessing cost is another major contributor to the poor plutonium economics.

If the reprocessing charge ($900/kg HM) includes both the reprocessing plant capital and operation and maintenance costs, thermal recycle will not be economically competitive with the current once-through system until about 55 years from now. Even if it is argued that the reprocessing plant has been built and the cost has been expended, the reprocessing operation and maintenance costs (half or $450/kg HM) will still make thermal recycle uneconomical over the next 32 years. Moreover, by the time that thermal recycle is economical, the current plants that countries want to keep running and be ready for the fateful day of economic recycling will have been retired anyway.

In the higher nuclear growth case (3.6 percent per year), economic thermal recycling will come only a little sooner—between 2020 and 2036 (Figure 3.8). Even the more optimistic date of 2020 is 27 years away. Existing plants will still have been retired or be near the end of their lives when thermal recycle becomes economical. In the meantime, the operation and maintenance costs alone will make reprocessing uneconomical.

Our projected uranium prices over time are considerably higher than the price scenarios used in the French government's latest study of "Reference Costs for Thermal Power Generation."[50] The French study assumed a yellowcake price of $20/lb U_3O_8 in the period 2000–2040 in one scenario and $30/lb in another. The latter assumes a significant revival of nuclear power worldwide. The study further says that the probability of uranium prices higher than $30/lb is "quasi nil." On the other hand, we have used in this study a

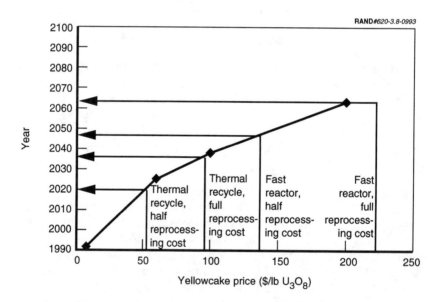

Figure 3.8—When Plutonium Thermal Recycle and Fast Reactor Become Economical (Higher Nuclear Growth Case)

[50]MacLachian (1993), pp. 1 and 12–14.

projected price range from $80 to $110/lb by the year 2040, or four times that used in the French study. This should assure that the dates we projected for economical thermal recycle and breeder are very conservative. The actual dates will likely be much later.

Countries have used resource conservation as a justification for their thermal recycle and breeder program. If a country is seriously concerned about running out of uranium, it might save the plutonium for fast reactors instead of using it in thermal recycle. Thus, the current rationale for thermal recycle in countries such as France, the United Kingdom, and Japan is weak, even from the point of view of conservation. Pursuing thermal recycle in industrialized countries gives an excuse to nuclear-aspiring countries such as North Korea to pursue reprocessing. They have argued that, if industrialized countries can pursue thermal recycle, they should be allowed to do the same. We recommend that thermal recycle activities worldwide be terminated or drastically slowed, since massive commercial recycling will not soon be needed. In Chapter Four, we will discuss how the economics of commercial thermal recycle affects the options of using military plutonium in commercial nuclear power plants as well.

If thermal recycle is uneconomical, fast reactors, having higher plant capital costs, are even more so. In the reference nuclear growth case for fast reactors, they are not expected to be economical until 2067, even if their power plant cost is only 10 percent higher than a LWR and the reprocessing cost dropped to half of the expected post-2000 cost. At 20 percent higher power plant capital cost and at the expected reprocessing cost, breeders are not expected to be economical until 2098. In the higher nuclear growth case, the breeders will be economical sometime between 2047 and 2063. Even the earliest date is more than 50 years away. (Recall that nuclear fission was discovered only about 50 years ago (in 1938) and the practicality of using it to harness the energy in the nuclei caused only a glimmer of hope in the scientists' eyes.) However, countries should not count on a new form of energy 50 years from now. We merely observe that 50 years is a long period of time in which other nuclear fission alternatives could be developed, not mentioning the possibilities of nonfission options such as fusion and solar power.

Strategies on plutonium use are about to enter a third phase. In the early days of nuclear power (1950s and 1960s), plutonium use was considered to be nuclear development's destiny. Then, in the 1970s and 1980s, plutonium use entered into a second phase. While its poor economics was gradually being recognized, proponents were still worried that there was no time to develop alternative proliferation-resistant systems or arrangements. They wanted to develop and deploy these proliferation-sensitive technologies as soon as possible to ensure against a uranium shortfall. Since the 1970s, scaled-back nuclear growth, discoveries of new uranium resources, rising plutonium reprocessing and fabrication costs, surplus fissile materials from weapons, and practically all other major factors have all been turning the tide even more against plutonium use.

Today, few plutonium-use proponents argue that it is economical to use thermal recycle and breeders now or that it will be in the near future. We argue that plutonium use should enter into the third phase in the 1990s and beyond. There is enough time to develop other fission alternatives and arrangements. In the third phase, plutonium use should be delayed indefinitely. Delay is countries' insurance policy against an irreversible plunge into a plutonium economy and the risks of nuclear proliferation. Moreover, even if the cost of plutonium use drops unexpectedly and its use becomes economical, the benefits of its use may not be large, because fuel cost is only a small component of electricity generating cost.[51] We should also note that if plutonium is needed unexpectedly, the construction of breeders could be accelerated. Although the average construction time of a nuclear power plant in the United States is 12 years, Japan manages to construct the same kind of plant in 5 years. Further, the U.S. experience in quickly constructing other types of plants in

---

[51]For example, with a yellowcake price of $10/lb and $70/SWU, the uranium and enrichment cost accounts for 2 percent and 3 percent of the busbar cost, respectively. Even at $100/lb, the busbar cost will increase by only 20 percent. In terms of delivered electricity price, which is typically about twice the busbar cost, the increase amounts to 10 percent. One way to formulate the proliferation risk and cost tradeoffs is to ask, "Are consumers willing to forgo plutonium use but to run a small risk of paying an x percent higher electricity bill for y years until plutonium use capability is reestablished?" In fact, the 20 percent increase is an overestimate because when the price reaches that high, it is economical to reduce the enrichment tail. In other words, we can use more units of enrichment but less uranium to get the same amount of enriched uranium fuel.

wartime emergencies indicates that the construction time can be shorter than 5 years.[52]

KEEPING ENOUGH PLUTONIUM FOR AN UNCERTAIN FUTURE

The strongest argument for keeping, instead of disposing of, military and civilian plutonium is to fuel future breeders when uranium resources are no longer adequate. Our calculation shows that breeders are not expected to be economical for 50 years. Still, how should countries prepare for a future in which uranium resources turn out to be much more costly to extract or nuclear generating capacity grows much faster than even our higher nuclear growth scenario predicts? It would be a good insurance policy to save enough plutonium to fuel breeders at any future date yet without incurring much proliferation risk. We will show that this is possible.

For use as fuel in nuclear reactors, weapon-grade plutonium is better than reactor-grade plutonium. Weapon-grade plutonium is 20 percent better than reactor-grade plutonium in fast reactors and 67 percent better in thermal reactors.[53] Since the plutonium is principally saved for breeders, the 20 percent figure is more pertinent here. In any case, if proliferation were not an issue and if only one type of plutonium could be saved, weapon-grade plutonium is preferred. On the other hand, the critical mass of reactor-grade plutonium is larger by about 40 percent and less attractive as weapon-making material. On this basis, there is relatively less proliferation risk in saving reactor-grade plutonium than in saving weapon-grade plutonium. However, the difference in neutronic value of the two types of plutonium for breeders and the difference in critical mass to nuclear aspirants are not large. These factors are overwhelmed by a third consideration—whereas both types of separated plutonium are radioactive, the radiation is not as intensive as in spent fuel. Therefore,

---

[52]Private communication with Bruno Augenstein, RAND, July 29, 1993.

[53]In other words, instead of using 1,390 kg of reactor-grade plutonium (weight of all plutonium isotopes that contain 960 kg of Puf) in a fast reactor, one can use 1,160 kg of weapon-grade plutonium (again, weight of all plutonium isotopes at weapon-grade isotopic composition). In LWRs, one can use 275 kg of weapon-grade plutonium instead of 460 kg of reactor-grade plutonium (which contains 320 kg of Puf).

both types of plutonium are not well protected against diversion for bomb-making. This third factor favors saving the plutonium that is still in a highly radioactive form. Saving spent fuel will accomplish the purpose.

There are four advantages to keeping spent fuel instead of separated plutonium. First, as mentioned above, spent fuel is highly radioactive, and it is difficult, costly, and time-consuming to extract plutonium from spent fuel. Therefore, spent fuel is considered to be proliferation-resistant, whereas separated plutonium is not. Second, on-site spent fuel storage and maintenance have already been paid for. The marginal cost of maintaining the storage beyond the nuclear power plant's operating life, if it is needed, is small. Thus, it is much cheaper to keep plutonium in spent fuel than in separated form. Third, separating civilian plutonium well in advance of need means that the reprocessing cost is paid too early and the storage cost must be paid until plutonium is needed. The argument about paying the reprocessing cost is not relevant for military plutonium from dismantled nuclear weapons, however, because that reprocessing has long since been paid for. Nevertheless, it is still expensive to store it. Fourth, the radioactivity buildup from Am-241 makes the handling of long-stored separated (reactor-grade) plutonium more difficult and costly (Figure 3.9). It would cost extra to remove Am-241 immediately before eventual plutonium use. In sum, storing spent fuel has little proliferation risk, is relatively inexpensive, allows the user to pay for reprocessing later and only when it is needed, and avoids the Am-241 buildup problem.

Countries may want to keep some spent fuel in storage so that they can fuel any reasonable program of breeder commercialization. The amount stored would depend on the quantity of plutonium required for breeder initial cores and the first few reloads and the rate of breeder capacity growth. The plutonium in about 30 annual discharges of LWR spent fuel will provide enough fuel for the initial core and the first three reloads of a Liquid Metal Fast Breeder Reactor (LMFBR). As to the determination of breeder growth rate, we examine the history of commercial nuclear power plants worldwide. The current power plants (thermal reactors) took 30 years to reach a nu-

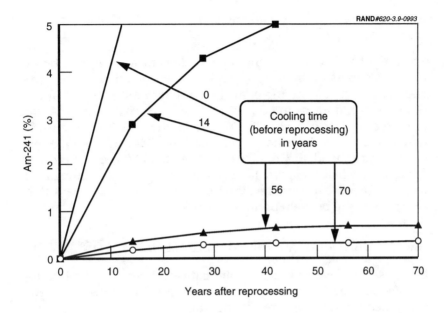

Figure 3.9—Am-241 Buildup in Reactor-Grade Plutonium

clear capacity of about 300 GWe by the year 1990. It is reasonable to assume that it will take the breeder the same amount of time to reach 300 GWe. To be more conservative, we assume that it will take only 20 years, instead of 30 years, for breeders to reach 300 GWe. Once that level is reached, the breeding of additional plutonium should be more than enough to sustain a subsequent growth of 1.8 percent a year, which is the expected growth rate for thermal reactors after the first 300 GWe (see Figure 3.4). Plutonium from spent fuel in thermal reactors can be used to supply much of the plutonium required for the startup cores and initial reloads of the first 300 GWe of breeders. We assume that all of the initial plutonium requirement will come from spent fuel: 9,000 GWe-yr (i.e., 300 GWe × 30 yr). This amounts to about 240,000 tonnes of spent fuel. The actual amount of spent fuel that needs to be kept at any time can be lower—6,000 GWe-yr (i.e., 300 GWe × 20 yr) or 160,000 tonnes of spent fuel—if we also save the spent fuel being accumulated during the 20 years of rapid breeder buildup.[54]

The amount of spent fuel that countries want to keep can be expressed as years of discharges of spent fuel at the then current nuclear capacity. For example, at the current world capacity of 340 GWe, the world needs to keep 17 to 26 years of spent fuel.[55] At the projected nuclear capacity of 474 GWe by the year 2010, the world needs to save 13 to 19 years of spent fuel—not a large amount. Every reactor has a mandatory on-site storage area for 30 years or more of spent fuel, which is more than adequate for storing enough for massive breeder commercialization. No additional spent fuel storage needs to be built for this purpose.

RECOMMENDED CIVILIAN NUCLEAR POLICY WORLDWIDE

In this section, we will discuss alternatives to sensitive fission options, especially for those plutonium-use proponents who are still concerned about running out of uranium.

There is little reason to pursue thermal recycle, but developing breeders has relatively better justification. Indeed, breeders can free countries from uranium import. But some countries, such as Japan, will still have to rely on imports for oil, foodstuffs, and many other critical products. It is difficult and expensive for a country to be self-sufficient in every way. Uranium and enrichment are available from many countries of different ideologies, and the possibility of a coordinated boycott against Japan or any other country is insignificant. Therefore, by not developing breeders, Japan risks not that other countries someday will not sell it uranium or enrichment but that the breeder is needed much sooner than anticipated. Japan worries that if other countries have developed it, they will use it to generate cheap

---

[54]Here, we made the conservative assumption that thermal reactor capacity will be merely 300 GWe for the future and that it will be replaced by breeders linearly over the next 20 years.

[55]If nuclear capacity does not grow, we want to save 26 years of spent fuel, on average, from every reactor (the higher value)—more from older or retired reactors to offset the fact that some reactors are young and sufficient spent fuel has not yet been saved from them. In reality, nuclear capacity is growing, and the average time in which to save spent fuel need to be slightly higher than 26 years.

energy and to gain a competitive edge in trade. Putting proliferation risk aside for the moment, there are penalties in following other countries' lead, especially when projects are costly and premature. Supersonic transportation is one such example. There are alternatives to the sensitive fission options that will assure Japan and others of adequate nuclear energy supply. Below, we offer a four-element program for the United States and other countries to pursue.

First, we recommend that the United States persuade other countries to terminate or drastically scale back reprocessing and plutonium use in thermal and fast reactors.[56] Any use of plutonium should be confined to demonstration and prototypes. Any sizable scale of commercial introduction in either nuclear or nonnuclear weapon states should be strongly discouraged. For example, Japan currently plans to use thermal recycle in two LWR plants by the mid-1990s, four plants by the late 1990s, and 12 LWRs by 2005. Although the United States would prefer to see Japan terminate reprocessing and thermal recycling completely, the United States should at least encourage Japan to scale down its thermal recycle program to only two plants by 2005. This plus the two breeder demonstrators—Joyo and Monju—would satisfy Japan's argument of needing to gain experience in reprocessing, thermal recycle, and breeders as an option against an unexpected shortfall of uranium. At the same time, such a scale-down would show other countries, including North Korea, Japan's concern about proliferation and its willingness to take action. The United States should also recommend that Japan cancel its planned construction of the two plutonium-fueled demonstrators—the Demonstration Advanced Thermal Reactor and the Demonstrator Fast Breeder Reactor. A scaled-back program in plutonium-use should include cancellation of JNFL's reprocessing and MOX fabrication plants at Rokkasho, currently under construction.[57]

---

[56]Neither is it necessary generally to extract the plutonium and uranium before permanent disposal of most types of spent fuel. One notable exception is the Magnox fuel used in U.K. reactors.

[57]The Rokkasho reprocessing facility has a designed capacity of 800 MTHM per year and can support thermal recycling (one-third MOX) in about 15 power plants. The Rokkasho fabrication facility has a designed capacity of 100 tonnes of MOX per year and can support about 12 plants (1 GWe each). These facilities are being built to support Japan's domestic thermal recycle program. If the program is scaled back as rec-

Japan and other countries that have a policy on storing separated plutonium for potential use can also help the cause of non-proliferation by storing instead natural uranium and plutonium-containing spent fuel.

Second, countries should rely longer on current once-through reactors, which are proliferation-resistant. They should pursue programs to improve uranium efficiency in current reactors. Higher burnup is one example. On the supply side, there is much more uranium available on earth than we have included in our calculations (Figures 3.7 and 3.8). Uranium is not a rare substance and has been estimated to be present at about 4 parts per million (ppm) of the earth's crust.[58] It is more abundant than silver, mercury, bismuth, and cadmium. But uranium prospectors do not purposely search for grades or types of uranium that cost more to extract than it is worth in the foreseeable future. Therefore, current projections are biased in that they tend to underestimate the ultimate amount of resources, especially those difficult to extract. As an example, twice in the past, once in the 1950s and once in the 1970s, when the price of uranium shot up (Figure 3.6), exploration efforts intensified and many uranium resources were found. Some were completely different types than expected.[59]

In the uranium resource estimate used in this report, we have not included most of the unconventional and by-product resources contained, for example, in marine phosphates, nonferrous ores, carbonates and black schists, and lignites. Of these possible sources, only marine phosphates have been roughly assessed. NEA/IAEA estimated that there are 9 million tons of uranium (U_3O_8) in phosphates in 14 countries but mostly in Morocco (8.5 million tons).[60] Morocco

---

ommended, these facilities need not be completed. Current facilities can support our recommended stretched-out program. The currently operating Tokai reprocessing facility has a designed capacity of 90 MTHM/yr and can support about two plants. In addition, Japan has long-term contracts with European reprocessors. The Tokai fabrication plants have a total designed capacity of 40 tonnes of MOX per year and can support about five plants.

[58]Glasstone and Sesonske (1967), p. 457. See also Benedict et al. (1981), pp. 234–236.

[59]The authors thank Gregory Jones for pointing out this fact to them on July 1, 1993.

[60]See NEA and IAEA (1992), pp. 27–28. As stated above, we expressed the amount of uranium resources in tons, instead of tonnes, because uranium price, especially in the United States, is customarily quoted in dollars per pound instead of per kilogram.

reported that its resources have an average grade of 120 ppm. The United States has already started to recover uranium from phosphates, although the amount is still small—about 1,300 tons of (U_3O_8) yellowcake in 1991. Nine million tons of (U_3O_8) yellowcake are enough to fuel 2,000 GWe of nuclear capacity, or 2,000 nuclear plants (1 GWe each), for their lifetimes. The current world nuclear capacity is 343 GWe, and 2,000 GWe amount to six times as much. If countries assess the resources better and explore extraction technologies, these unconventional resources may provide a cushion to any unexpected shortfall of uranium. Also, recall that there is tremendous amount of uranium in seawater, which can support current types of thermal reactors indefinitely. Although uranium from seawater will be much more expensive to extract than the conventional uranium resources, it might still serve as an insurance against running out.

Third, countries should, over the next 50 years, develop proliferation-resistant reactors that have very low uranium requirements. For example, researchers at Brookhaven National Laboratory identified a reactor concept—the fast-mixed spectrum reactor (FMSR).[61] It uses both slow and fast neutrons to generate power and breed plutonium. It is designed to operate on a once-through-and-store fuel cycle, with fuel staying in the reactor for as long as 17 years and no fuel reprocessing. It is a version of the once-through fast reactor. The core and initial reloads will consist of enriched uranium of about 10 percent, well below the weapon-usable grade of 20 percent. Only natural or depleted uranium, but no additional enriched uranium, is required for reload after the first core and transition cycles. The plutonium burned in the reactor is produced in situ by neutron capture. Such a reactor will consume less than one-fifteenth of the uranium that current types of nuclear power plants do. In other words, if we have enough uranium to support the nuclear power of a given capacity for merely 30 years now, using FMSRs instead can support the same capacity for more than 450 years. This is a very long time by any planner's standard. Thus, there are promising concepts that can greatly extend the uranium resources and are proliferation-resistant. Countries should be exploring these alternatives.

---

[61]Kouts and Kato (1978). See also Chow (1981), pp. 31–38.

Fourth, countries should establish international controls for potential plutonium use. Although nonproliferation concerns would favor not using separated plutonium at all, countries should explore ways to minimize the risks of employing sensitive systems in the event that they are used after all. How should the world deal with the situation where France, the United Kingdom, the FSRs, Japan, and others are not willing to mothball their reprocessing facilities and the breeder demonstrators, no matter how hard the United States and other countries try to persuade them? The world needs to recognize the difficulty in handling plutonium use at the onset: As long as some countries pursue plutonium use, other countries will argue that they can, too. It is very difficult to separate countries into two groups— one that can pursue plutonium use and one that cannot. On the other hand, if a line must be drawn between who can and who cannot pursue plutonium use, the most logical line is between nuclear weapon states and nonnuclear weapon states. That distinction, however troublesome or inequitable, has already been made in NPT and in IAEA, and will continue to be the one used until and unless total nuclear disarmament is adopted and implemented. The most serious problem in plutonium use is the existence of separated plutonium in nonnuclear weapon states. A nuclear aspiring nation can seize the separated plutonium, and it will be too late for IAEA or the United Nations to take effective counteractions. Keeping nuclear aspirants from obtaining weapon-usable fissile materials has long been the cornerstone of the international nuclear control regime. Allowing nonnuclear weapon states to possess separated plutonium is in direct violation of this principle and will render the current nuclear control regime ineffective.

The world needs to develop an international nuclear fuel cycle regime such that sensitive facilities and activities that use plutonium, as well as uranium enrichment, are confined within nuclear weapon states, but with any benefits being shared by all countries. Any exceptions will be transitional and will be eliminated when international arrangements are in place. Exceptions will be made only when mothballing or moving existing plants to a nuclear weapon state or stopping projects well under construction will create a severe financial hardship that cannot be compensated. Plants in Japan and the Netherlands may qualify for such exemption. No exception, however, should be made for plants still in the planning stage.

To make the international arrangement more acceptable to nonnuclear weapon states, nuclear weapon states need to accommodate them in several areas. First, if countries such as Japan cannot pursue breeders, nuclear weapon states should not be allowed to use nuclear energy to gain competitive advantages. If breeders are ever used commercially in nuclear weapon states and if electricity from breeders turns out to be cheaper than that from LWR (OT), a tax could be levied against the breeders and the proceeds used to compensate nonnuclear weapon states that could have developed and employed breeders. We have not worked out the details and mechanisms of taxes and rebates, but the principle is that the nuclear electricity cost to them would be the same as if they had developed and deployed the breeders.[62] Second, international uranium fuel banks could be established to stock natural and perhaps even some enriched (from blended-down HEU) uranium. These banks could be located in nonnuclear weapon states, especially in countries that have forgone their active breeder programs. Moreover, nonnuclear weapon states, if they are members of this international cooperative, would have first claim on uranium fuels in the fuel banks in the event of emergency. Third, nonnuclear weapon states could be shareholders in international consortia that develop and even own breeders, although they could not technically participate in their system development. Fourth, nonnuclear weapon states could pursue research, development, and production of components of sensitive systems, as long as these components themselves were not sensitive. (Such components include system generators, heat exchangers, valves, temperature and other sensors, gauges, and many others.) Current reactor manufacturers use components supplied by subcontractors in other countries. The same arrangements, or even more subcontracting, could hold for sensitive systems in the future. Fifth, both nuclear and nonnuclear weapon states should be encouraged to develop proliferation-resistant

---

[62]The calculation will include the amortization of breeder research and development cost that these countries would have to make in the future. Moreover, in the unlikely event that the electricity cost from nuclear fission under our proposed international arrangement turns out higher, the higher cost will be shared by all countries.

reactors such as the fast-mixed spectrum reactor (already discussed) and advanced reactors based on U-233. The objective of a U-233-based converter is that, in the event breeders are needed, plutonium use can be confined to nuclear weapon states and plutonium will be used to turn Th-232 into U-233. Nonnuclear weapon states' spent fuel could be reprocessed in nuclear weapon states. Low-enriched U-233 could be returned to nonnuclear weapon states for use in their U-233-based reactors.[63]

An international regime to govern plutonium use would insure against an unexpected shortfall of uranium. Then, in the likely event that plutonium use will not be needed for the next 50 years, the world would not have plunged into a plutonium economy and unnecessarily incurred the nuclear proliferation risk.

In sum, since plutonium use will not be economical for many years, countries now have enough time to explore other alternatives, especially fission alternatives, and plutonium use can be postponed indefinitely.

---

[63]Feiveson and Taylor (1976), pp. 14–18 and 46–48.

Chapter Four

DEALING WITH PLUTONIUM FROM DISMANTLED NUCLEAR WEAPONS

In Chapter Three, we recommended a four-element program for a proliferation-resistant civilian nuclear future; we now examine the technical feasibility, cost, and proliferation effect of various alternatives in dealing with military plutonium. The chosen option should accomplish two criteria. First, it should prevent FSRs as much as possible from fashioning these high-grade weapon materials back into nuclear weapons or selling them to nuclear-weapon aspiring countries or groups. Second, any actions taken should at least not hinder the movements toward a proliferation-resistant nuclear future.

TECHNICAL FEASIBILITY OF ALTERNATIVE PLUTONIUM-HANDLING SCHEMES

Since the end of the Ford administration, the United States has adopted a policy of deferring the use of plutonium in civilian nuclear power plants. Thus, the plutonium from U.S. dismantled nuclear weapons should be placed in storage in the near term. On the other hand, it is uncertain how plutonium from FSR weapons will be used or managed. FSR plutonium is being considered for use in thermal nuclear reactors and fast reactors.[1] All major industrialized countries

---

[1]It was mentioned earlier in the report that most current nuclear power plants are thermal nuclear reactors or power plants, designed to use slow (thermal) neutrons to maintain the nuclear chain reactions. On the other hand, fast reactors rely on fast neutrons and have more potential to breed more fuel from fertile material than is

have shown interest in helping FSRs deal with their plutonium. Their motivations are nonproliferation and economic benefits. France, the United Kingdom, and Japan have long considered plutonium use in their reactors a key element of their nuclear programs, in spite of deteriorating plutonium economics.[2] Instead of postponing pluto- nium use, they all see the use of military plutonium in thermal and/or fast reactors as helping them spread the high fixed cost of fabrication plants for MOX or breeder fuels. They also hope that us- ing nuclear power plants to get rid of nuclear weapon materials would help rally public support in using plutonium for civilian pur- poses. Finally, profitmaking plays a role in these countries' pursuit of FSR military plutonium. They think that the United States, if not the FSRs, might be willing to pay for getting rid of FSR military ma- terials.

Some argue that burning plutonium in reactors is not really getting rid of plutonium, especially if plutonium will eventually be recovered from spent fuel. Those who advocate plutonium use in reactors counter by noting that the power-plant-generated plutonium will be rich in Pu-240 and Pu-241 and is less desirable for use in nuclear weapons. Typical reactor-grade plutonium contains 24 percent Pu- 240 and 11.5 percent Pu-241, whereas typical weapon-grade pluto- nium contains merely 6 percent Pu-240 and 0.5 percent Pu-241. The problem with Pu-240 is that it is not as fissionable as Pu-239 by fast neutrons for nuclear explosives, and more plutonium will be re- quired to make a nuclear bomb.[3] Moreover, its high rate of neutron emission from spontaneous fission can cause preinitiation and results in a statistical distribution in yield between low and high values Pu-241 has a relatively much shorter half-life of 14.4 years than other plutonium isotopes, and it beta-decays into Am-241.[4] Am-241's alpha-decay into Np-237 generates a considerable amount

---

consumed. The Liquid Metal Fast Breeder Reactor is the most popular type of fast reactor and is being developed in France, the FSRs, Japan, and other countries.

[2]Germany has recently canceled its SNR-300 breeder demonstrator and reprocessing. But many in the nuclear establishment still strive to revive these activities.

[3]The bare critical mass of Pu-239 is 10 kg and that of Pu-240 is about 40 kg (Selden, n.d.).

[4]Pu-241 will also alpha-decay into U-237, which then alpha-decays into Np-237 with a considerable amount of gamma radiation. But since the half-life of U-237 is only 6.75 days, the radiation would decline quickly (Benedict et al., 1981, pp. 137–142).

of gamma radiation and heat, which is more difficult to shield and complicates plutonium handling, fabrication, and bomb design.[5] These weaknesses are not serious, and nuclear aspirants would probably not be deterred from making bombs because the only bomb materials accessible to them were less than perfect. After all, the critical mass for reactor-grade plutonium is merely 6.6 kg, or only 40 percent more than that for weapon-grade plutonium. Although a bomb made of reactor-grade plutonium has uncertain yields, there can still be high confidence that the bomb will be in the kiloton range and will be much more destructive than conventional explosives. As to the gamma radiation and heat from Am-241, nuclear aspirants could separate the Am-241 from plutonium or take the radiation and heat into consideration when they design the handling facilities and the bomb.

Another problem relates to using military plutonium in civilian reactors in nonnuclear weapon states. It could make weapon-grade plutonium accessible to nuclear-aspiring countries. Even if certain destabilizing countries could be excluded, a country's political system and ideology can change. Recall Iran's drastic political changes since the Shah in the late 1970s. Even Japan, which many are confident will remain stable far into the future, had at one point recently chosen to keep open the option of developing nuclear weapons. The option of restricting the use of military plutonium to only industrialized countries such as Japan would make sense only if such use is highly cost-effective and can greatly reduce countries' vulnerability to energy supply disruption. On the other hand, if such use engenders political outcries and economic penalties, the option is unattractive and will merely further the complaints of the third world that the West consistently excludes them from profitable ventures. Below, we analyze the economics of using military plutonium in LWR (PM), LWR (FM), and fast reactors.[6] We also examine plutonium storage and disposal costs.

---

[5]Pu-239 generates 1.9 mW/gm, whereas Am-241 generates 114 mW/gm (NEA, 1989, p. 24).

[6]As defined earlier, LWR (PM) uses plutonium-bearing MOX fuel for a one-third (or partial) load in light water reactors. LWR (FM) uses 100 percent or full MOX fuel.

This study does not provide cost estimates of the following options: using plutonium for space flights; transmuting the plutonium; detonating nuclear warheads underground, either for electricity generation or just for their elimination; and sending the plutonium to the sun. The technologies for these options and their effects on the environment are too uncertain at this time.

The ORION program for robotics space flight is considered by some to be a real technical possibility, as reported at a recent conference.[7] Each such flight is likely to consume a number of tons of plutonium, which is only a small fraction of the weapon-grade plutonium available from FSRs.

Another scheme for eliminating plutonium is to transmute it to nonweapon-usable elements. Here, the basic technology is well in hand but the plan is not ready for a full test. Major strides, partially fueled by work under the Strategic Defense Initiative, have made the technology more tractable, and the development periods can be shortened. The key question is the cost of plutonium elimination and the remaining cost of disposing of the transmuted elements and other radioactive waste.

A concept to turn nuclear warheads into plowshares is to explode the warheads in a specially constructed underground chamber that contains heat transfer systems leading to electricity generators.[8] The concept has two attractions. It can use weapons nearly "as is" in weapon form and can use up both the fissile and fusion components of nuclear weapons. Such a system will not be available soon, and keeping nuclear warheads intact would run opposite to the intent of START 1 and 2.

Scientists at one of the two Russian nuclear weapon design laboratories, Arzamas-16, proposed destroying the plutonium portions of nuclear warheads. They envision using a 100-kiloton nuclear explosion at 1-km underground to destroy 5,000 of these plutonium components at a time.[9] Following their assumption that these compo-

---

[7]Augenstein (1991).

[8]Private communication with Bruno Augenstein, RAND, on August 1, 1993.

[9]Trutnev and Chernyshev (1992), pp. 31–32.

nents contain 20 tons of weapon-grade plutonium, it would take only six such explosions to get rid of the FSR 110 tonnes (or 120 tons) of surplus plutonium. In each explosion, the plutonium would be dispersed relatively uniformly throughout 30,000 tons of vitrified rock. The 0.0007 gram of plutonium per gram of melt would have a radioactivity of only 5×10^{-5} curies per gram. By comparison, uranium ore has an activity of 4×10^{-10} and uranium concentrate, 4×10^{-7}.[10] The problem with this scheme is the technical uncertainties and long-term effects on the environment, to say nothing about local opposition to the selected site for this instant repository.

Another often-mentioned scheme is to shoot the plutonium to the sun. Theodore Taylor envisioned a scheme of using heavy boosters such as the U.S. Saturn V or the Russian Energia to send a few tonnes of plutonium at a time into high earth orbit.[11] The payload would then be tugged into the solar orbit by solar power and decelerated until it dropped into the sun. Taylor estimated that it would take several tens of launches to dispose of 200 tonnes of plutonium, which coincides with our estimate of the surplus weapon-grade plutonium in the United States and FSRs. Moreover, he believed that the plutonium package can be designed to survive an accident involving the explosion of the booster, reentry into the atmosphere, and the impact on the earth surface. He estimated the cost at $10 billion. We will show below that this cost is about five times as much as burning weapon-grade plutonium in light water reactors or storing it for 20 years. Another disadvantage of this scheme is the uncertainty of whether the plutonium package would actually survive a launch accident. The advantages of this scheme are that it will not encourage plutonium use and it avoids the environmental and political difficulties of finding a repository for it on the earth.

PLUTONIUM BURNING COSTS

Even if military plutonium can be obtained free, its use is not necessarily economical. The extra cost of handling plutonium could outweigh the savings from using less uranium and enrichment. One key

---

[10]von Hippel (1992), p. 32. See also his comments on this scheme in Appendix J, pp. J-21 to J-22.

[11]The discussion in this paragraph is based on von Hippel (1992), pp. J-21 to J-22.

extra cost, which was discussed in Chapter Three, is the cost of fabricating MOX fuel over and above that of uranium dioxide fuel. We applied the electricity cost model and the data in Table 3.1 to calculate whether military plutonium has a positive value. Basically, we calculated the total cost for generating a given amount of electricity in a year at the current price of $10/lb U_3O_8. In the case of using military plutonium, the plutonium is assumed to be free. If the total electricity costs using plutonium are less than not using plutonium, dividing the savings by the amount of plutonium involved will give us the value of plutonium in dollars per kilogram of plutonium. On the other hand, if using plutonium costs more, the same calculation will lead to the cost of eliminating military plutonium.

We found that military plutonium is a liability rather than an asset. Whether in LWR (PM), LWR (FM), or in fast reactors and whether or not the spent fuel is reprocessed, using plutonium will make electricity generation more costly than the current LWR (OT). In other words, it will cost money to burn or eliminate military plutonium in nuclear reactors. We estimate that it will cost $7,600 to eliminate 1 kg of weapon-grade plutonium in the current LWR. This assumes that the spent fuel from such a reactor is not reprocessed. The poor economics of reprocessing would make the cost to eliminate weapon-grade plutonium much higher, whether in LWR (PM), LWR (FM), or in fast reactors. For example, the cost to eliminate 1 kg of weapon-grade plutonium in LWR (PM) jumps to $62,000, an eightfold increase, if the cost of reprocessing the spent fuel is also included in the calculation. If one uses the weapon-grade plutonium in current fast reactor demonstrators without reprocessing,[12] the cost of eliminating each kilogram of weapon-grade plutonium is $18,000. If one uses LWR (FM) without reprocessing to eliminate military plutonium and if a LWR (FM) does not cost more than the current LWR, it will still cost $5,600 to burn 1 kg of weapon-grade plutonium.

The nuclear industries in the West seem to think that they can charge FSRs a fee for getting rid of their weapon-grade plutonium. But the FSRs believe that they can sell plutonium to others for a profit. This

---

[12]Here, we do not include the higher power plant capital cost of building a fast reactor, because the demonstrator has already been built and the capital cost differential, as well as the capital cost, is sunk.

diametrically opposite perception makes any agreement in dealing with military plutonium much more difficult than with HEU, which everyone agrees is a valuable asset.

PLUTONIUM STORAGE COST

The total cost for plutonium storage comprises two principal components: (1) capital costs of the storage facility and its equipment and (2) operating costs. Bloomster et al. estimated that the capital cost for storing 50 tonnes of plutonium would consist of $180 million for the building and $74 million for the equipment.[13] We scaled these costs for a facility to store the 100 tonnes of U.S. plutonium first.[14] Then, we scaled the capital costs for storing both 100 tonnes of U.S. plutonium and 100 tonnes of FSR plutonium. Comparing these figures, we can determine the marginal cost of storing FSR plutonium. The space required for the plutonium containers is expected to be proportional to the amount of plutonium to be stored. On the other hand, space required to process and monitor twice as much plutonium would not double the space, because of economies of scale. Using a scaling rule derived from civilian reactor expenditures, we expect that, as the amount of plutonium doubled to 100 tonnes, the space for handling would only be increased by 1.5 times. Similarly, as the amount of plutonium to be stored quadrupled to 200 tonnes, the space will only be doubled. Consequently, we determined that the capital cost for a 100 tonne storage facility would be $310 million for the building and $110 million for equipment. The costs corresponding to a 200 tonne facility would be $540 million and

---

[13]Bloomster et al. (1990, p. 12) started with the GESMO estimate of $215 per square foot of floor area (in 1973 dollars) (NRC, 1976). They translated it to $6,700 per square meter in 1990 dollars. They assumed that each container holds 4 kg of plutonium and occupies one square meter of space. Thus, to store 50 tonnes of plutonium, one would need 12,500 containers and 12,500 square meters of space. They further assumed that the space for equipment and other purposes would take up another 12,500 square meters. The capital cost for the building would be $170 million in 1990 dollars. They estimated the capital cost for equipment, such as security and verification instruments, cranes, and equipment for safety and environmental analysis, to be another $70 million. We converted these figures to 1992 dollars: $180 million and $74 million, respectively.

[14]In this section, we use 100 tonnes each as the U.S. and FSR available inventory of plutonium, which is not too different from the figures of 91 and 109 tonnes in Table 2.1.

$150 million.   Thus, the marginal costs for storing FSR plutonium (the second 100 tonnes) would be $230 million and $40 million.   On a capital cost per kilogram basis, the 200 tonne facility costs $3,450/kg and the marginal cost for the FSR plutonium is $2,700/kg (see Table 4.1).

According to Bloomster et al., a plutonium storage facility built for storing 50 tonnes of plutonium will require 200 on-site staff members, with 150 physical security staff and 50 staff involved in the operation, maintenance, and administration of the facility. Assuming $110,000 per fully burdened staff-year, they estimated an annual salary-related cost of $22 million and an additional $6 million in other expenses.

A facility four times this size (built to accommodate 100 tonnes of FSR plutonium and 100 tonnes of U.S. plutonium) should not require four times the operating staff. Using the same scaling rule as above, we found the following operating costs: 50 tonnes of plutonium facility, $28 million per year; 100 tonnes of plutonium facility, $42 million per year; 200 tonnes of plutonium facility, $56 million per year.

The marginal operating cost for storing the extra 100 tonnes of FSR plutonium would be $14 million per year (i.e., $56 million minus $42 million), or $140/kg-yr.

Table 4.1

Annual Storage Cost per Kilogram of Weapon-Grade Plutonium

	Cost to Store U.S. and FSR Pu (200 Tonnes)	Marginal Cost to Store FSR Pu (100 Tonnes)
Building	$540 million	$230 million
Equipment	$150 million	$40 million
Capital cost/kg	$3,450	$2,700
Annual capital recovery (30-year plant life) @ 10 percent	$370/kg-yr	$290/kg-yr
Operating expenses	$280/kg-yr	$140/kg-yr
Total @10 percent	$650/kg-yr	$430/kg-yr

By comparison, Carter and Cote (1993) quoted a lower capital and operating cost. They stated that the cost for a single storage facility capable of storing 100 tonnes of fissile material in metal form is likely to be about $250 million for construction and equipment and $10 million annually for operation. Our corresponding estimates are $420 million and $42 million for a 100 tonne storage facility. Their figures would translate into an annual charge of $370 for each kilogram of plutonium stored, whereas ours translate to $870/kg-yr—about twice as much. This $870/kg-yr figure can also be used as the annual storage cost for the 100 tonnes of U.S. plutonium. The marginal capital cost of the building and equipment to store 100 tonnes of FSR plutonium in the United States is $270 million and the annual operating cost, $14 million. These figures lead to a charge of $430/kg-yr (see Table 4.1).[15] The cost of 20-year storage would be $3,800 per kilogram at a 10 percent discount rate.

PLUTONIUM DISPOSAL COST

A plutonium disposal facility would differ from a plutonium storage facility in that the disposal facility would likely be underground and require less manpower for maintenance and security.

Separated plutonium could be disposed of with or without mixing it with other wastes or spent fuel. The direct disposal of separated plutonium is a possibility, according to PNL. PNL estimated, in 1990,

---

[15]This $430/kg-yr figure assumes that the 200 tonne storage facility for U.S. and FSR plutonium operates at full capacity from the start and that only the marginal cost for the second 100 tonne capacity is attributable to FSR plutonium. In practice, as the nuclear weapons are dismantled, the amount of plutonium placed into the facility may build up gradually. One way to reduce the capital cost charge is to start with a smaller plant and expand its storage capacity in steps commensurate with the plutonium buildup. Even using this approach, one would expect the storage charge on a per kilogram and per year basis to be higher during the initial years, because the charge is higher for a smaller plant However, we have not adjusted the $430/kg-yr figure upward for two reasons. First, our estimate is already about twice as high as that of Carter and Cote. Second, even if the actual figure is twice as high as ours, our recommendation does not change. We recommend that the United States offer to purchase FSR plutonium. Once it is taken out of the FSRs, whether to store it, to use it as fuel in reactors, or to dispose of it should be more a nonproliferation issue than an economic one. Doubling the storage cost would increase the discounted cost of storing FSR plutonium in the United States for 20 years by about $400 million, not a large sum by national security standards.

that the direct disposal of large quantities of plutonium as PuO_2 in a geologic repository such as WIPP (Waste Isolation Pilot Plant) may be feasible within the next five to fifteen years. PNL estimated that cost at $9,000 per kilogram of plutonium. However, the cost could be drastically reduced by waiting until the waste or spent fuel is ready for disposal and then mixing and disposing of plutonium and spent fuel in the same operation. This way, the plutonium disposal cost is only the marginal cost of the combined disposal. PNL estimated the marginal cost at only $1,000 per kg if the separated plutonium is mixed with defense high-level waste. The disposal, however, might not be feasible for 20 years, because a suitable repository would not be available before the year 2010.

COMPARISON OF PLUTONIUM OPTIONS

Previous sections provide the basic economic comparisons of the options of burning, storing, and disposing of weapon-grade plutonium. These data are summarized in Figure 4.1. We examined five options for dealing with weapon-grade plutonium. The first is to use or burn[16] the plutonium in existing fast reactor demonstrators without reprocessing. The second option is to use it in light water reactors fueled with one-third or partial plutonium-bearing MOX fuel without reprocessing—LWR (PM, w/o R). The third option is to use the plutonium in light water reactors fully fueled with MOX without reprocessing—LWR (FM, w/o R). The fourth is to store plutonium for, say, 20 years. The fifth is to dispose[17] of the plutonium by mixing it with waste or spent fuel when the waste or spent fuel is being prepared for final disposal. None of these options creates any commercial value for weapon-grade plutonium.

---

[16]Although "burn plutonium" and "use plutonium" are used interchangeably here, other reports may use the former to emphasize the fact that plutonium use is uneconomical.

[17]Plutonium disposal means that plutonium is placed in a repository, not in a reactor.

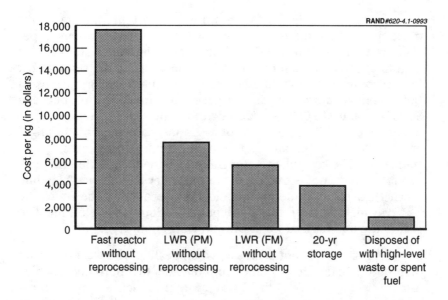

Figure 4.1—Cost of Storing or Disposing of FSR Weapon-Grade Plutonium

In the first three options, even if the weapon-grade plutonium is free, the extra cost in handling the highly radioactive and toxic plutonium outweighs the savings from using less uranium and enrichment services. Burning plutonium in fast reactors actually has a net cost of $18,000/kg; in LWR (PM), $7,600/kg; and in LWR (FM), $5,600/kg. The storage cost for 20 years is $3,800/kg. The cheapest way to dispose of plutonium is to mix it with waste or spent fuel being prepared for final disposal. The marginal cost for this approach is $1,000/kg. But the U.S. repository will not be ready for operation until 2010 and neither FSRs nor other countries have such repositories. In the meantime, it would be expensive to mix weapon-grade plutonium with the existing high-level waste or spent fuel if, as in the United States, the HLW or spent fuel is kept in sealed tanks or in its original cladding. On the other hand, the United States and FSRs should want to have disposal of separated plutonium as a future option. Making this option available is an added motivation to speed up the process in developing repositories in the United States and FSRs for both military and civilian waste.

Thus, to adopt the disposal option, interim plutonium storage cost must also be factored in. Even in the three burning options, especially in currently unavailable LWR (FM)s, plutonium storage cost might have to be paid for a few to 10 years. Taking the storage cost into account, the cost differences among the burning options in LWRs and the store-now-and-dispose-later option are all between $4,000/kg and $10,000/kg. Although the difference in total cost for handling the FSR 110 tonnes of weapon-grade plutonium might amount to $660 million, that is not extremely large by national standards. The key policy factor should still be the proliferation risk in each option, not economics. On the other hand, since blending down HEU resolves the proliferation risk, economics becomes the key consideration for the HEU policy. Comparison of the five plutonium options also yields the following comments.

First, reprocessing is not necessary when dealing with weapon-grade plutonium. In fact, if spent fuel from reactors that burn military plutonium is reprocessed, the extra cost, not incurred with once-through, would make the cost of getting rid of military plutonium much higher.[18] A support of plutonium burning does not imply a support of reprocessing. The former activity transfers separated plutonium into spent fuel; the latter does just the opposite, taking plutonium out of spent fuel.

Second, although eliminating weapon-grade plutonium through a LWR (FM, w/o R) is the least costly plutonium-burning option, such burners do not exist today. Their development and licensing will take at least several years. Since each LWR (FM) will consume about 0.9 tonne weapon-grade plutonium a year, and since these plants would be available only, say, 5 years from now,[19] 24 plants would be needed to get rid of all the FSR surplus plutonium by 2003 or 12 plants to get rid of it by 2008. On the other hand, nuclear plant life-

---

[18]Reprocessing would increase the cost of eliminating weapon-grade plutonium by LWR (PM) from $7,600/kg to $62,000/kg, and by fast reactors from $18,000/kg to $39,000/kg. In this calculation, the reprocessing cost for LWR spent fuel is $900/kg HM and that for fast reactor fuel is $1,080/kg HM. These costs are conservative; current charges are in the $1,400 to 1,800/kg HM range.

[19]This assumes that some existing plants or plants under construction can be converted for this purpose. Otherwise, it will take even longer to construct new plutonium burners.

times are expected to be 30 years. Countries would have to stretch the schedule in eliminating weapon-grade plutonium or they would have to expect that these plants will eventually be fueled by reactor-grade plutonium. The latter alternative leaves two policy options. If countries decide to close down or scale back their reprocessing activities, these plutonium-burning plants can be used to eliminate or reduce the existing reactor-grade plutonium inventory. In the second case, reprocessing could continue and, in fact, would get a boost from these plutonium-burning plants. These plants would provide an added demand for the plutonium being separated by reprocessing. Then, attempts to eliminate separated plutonium might promote plutonium separation in the long run. The proliferation risk of separated plutonium in the economy makes this second case unfavorable, at least until the economic benefits are very compelling. But this is not the situation at present and there is no sign that it will be in the near future. Hence, it seems inadvisable to enter into a situation that is much more likely to generate proliferation risk than economic benefits.

Third, regardless of which option to take, more plutonium storage is needed somewhere in the world.[20] LWR (FM)s are not available now. Plutonium from dismantled nuclear weapons, if it were to be burned in LWR (FM)s, would have to be stored in the meantime. Even after plutonium is used in LWR (FM)s, LWR (PM)s, or fast reactor demonstrators, storage for plutonium will still be needed to hold inventory. Plutonium storage will also be needed in cases where plutonium use of any kind is prohibited because of difficulties in agreeing which countries can use plutonium and which cannot. Finally, if permanent disposal of separated plutonium is the chosen option, plutonium storage will be needed until a repository to accept waste and spent fuel is opened. We indicated above that the storage cost can be $870/kg a year for the U.S. plutonium, which is lower than the current charge in Europe of $1,000 to $2,000 per kilogram of plutonium.[21] The marginal cost of storing FSR plutonium in the United States can be even lower. Since the United States needs to have facilities for its own plutonium anyway, the cost of storing FSR plutonium in the United States can be treated as marginal or added

---

[20]Carter and Cote (1993), p. 125, came to a similar conclusion.

[21]NEA (1989), p. 58.

cost to the U.S. storage facilities. Many of the common facilities and security systems have already been charged against U.S. plutonium. The international community should actively support a program to develop and construct much cheaper plutonium storage facilities.[22] The plutonium storage locations and whether they should be under unilateral, bilateral, or IAEA safeguards depend, however, on the actual option chosen. If, in addition to their HEU, FSRs are willing to sell their weapon-grade plutonium to the United States, much less storage would be needed in FSRs. On the other hand, additional storage for plutonium and HEU would be needed in the United States, especially if it was decided to withhold the release of plutonium and blended-down HEU to minimize market disruption. Finally, if plutonium from both the United States and FSRs is to be burned in France and the United Kingdom, less storage space would be needed in the United States. These are, however, secondary issues until FSRs decide whether to sell their plutonium.

RECOMMENDED U.S. POLICY TOWARD WEAPON-GRADE PLUTONIUM FROM FSRs

The United States should make FSR weapon-grade plutonium as inaccessible to FSRs as possible so that it is difficult to make nuclear weapons with it. The best way to accomplish this is to remove plutonium from FSRs. A second-best option is to encourage burning of FSR weapon-grade plutonium but with a corresponding reduction in civilian plutonium reprocessing.

Removing Weapon-Grade Plutonium from FSRs

A good way to remove weapon-usable plutonium from FSRs is to buy it. The FSRs are willing to sell their HEU and it is possible that they may be willing to sell their plutonium also. The United States may have to negotiate with the independent republics as an entity, and with the four nuclear republics, especially Ukraine, individually. For both strategic and economic reasons, the non-Russian nuclear re-

---

[22]The storage facilities can also be used to store HEU while it is waiting to be blended down. It may be more convenient and cheaper to keep some HEU in storage collocated with blend-down facilities.

publics may prefer to sell their plutonium to the United States than to sell or simply transfer the plutonium to Russia. The United States can argue that military plutonium currently has no economic value and will be costly to store.[23] We recommend that the United States offer the FSRs, say, $10,000 for each kilogram of weapon-grade plutonium.[24] The payment could be made as plutonium is received. We estimate that the FSRs will have 110 tonnes of surplus weapon-grade plutonium by 2003 (Table 2.1). Therefore, the FSRs could receive $1.1 billion (undiscounted) over the next 10 years. Once plutonium is purchased and removed from FSRs, it could be stored or burned, depending on what bargain we could strike with FSRs and our allies. If it is stored, the United States is the preferred location, but storage in the United Kingdom or France would be acceptable, too. If it is burned, that could be accomplished in the already available plutonium-bearing fabrication facilities and suitable nuclear power plants in the United Kingdom and France. But the United States should negotiate with them to reduce the amount of plutonium to be recovered from spent fuel. This way, weapon-grade plutonium could be eliminated while discouraging plutonium separation from spent fuel. Thus, the United States should seek money from the United Kingdom and France to purchase FSR weapon-grade plutonium and/or should ask them to burn this plutonium without charging the United States or FSRs a fee.

The offering price we propose for the weapon-grade plutonium represents its value in nuclear power generation, when the yellowcake price is $39/lb—the current price being around $10/lb. As shown in Figure 4.2, the plutonium is most valuable when it is used in LWR (FM, w/o R), and that is the price we should offer to FSRs. Although we do not expect the price of uranium to reach $39/lb any time soon, others might. If the FSRs believe that the yellowcake price will rise beyond $39/lb soon and that the United States is not paying enough for it, the United States might consider offering to reestimate plutonium's value every five years for the next 20 years. For example, if the average spot yellowcake price over the prior five years was

---

[23]By 2003, it will cost the FSRs about $870/kg-yr or $96 million a year to store their 110 tonnes of surplus military plutonium.

[24]The United States should, however, be flexible in the purchase price.

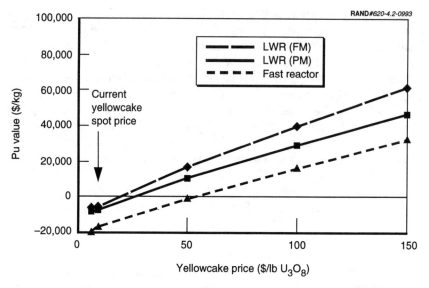

Figure 4.2—Value of Weapon-Grade Plutonium in Civilian Reactors (no reprocessing)

$60/lb, the United States would pay $20,000/kg for all the weapon-grade plutonium delivered during that five-year period under the agreement. If yellowcake price was $82/lb, the United States would pay $30,000/kg, and so on, up to a cap of $40,000/kg at $104/lb of yellowcake price.[25]

Disposing of or Burning FSR Weapon-Grade Plutonium in FSRs

Even after the best efforts of the United States and others, the FSRs might still refuse to let their weapon-grade plutonium leave their countries, even though it has no economic value and will cost them

---

[25]At other yellowcake prices, the plutonium value will take the interpolated figure. These are the plutonium values at the corresponding uranium prices, using this study's projected input values.   As an alternative, the plutonium value can be determined from then-current yellowcake price, reprocessing cost, fabrication cost, etc. We do not recommend this alternative, because both sides may not agree to the data at the time of reevaluation.

money to manage. Then, a second option would be to dispose of it in the FSRs. Unfortunately, like other countries, the FSRs might not have a repository for the next 10 to 20 years. Burning weapon-grade plutonium in the FSRs also faces problems. For at least the next several years, the FSRs will lack the needed fabrication facilities and appropriate nuclear reactors to eliminate the weapon-grade plutonium released from their dismantled weapons.

Nonnuclear weapon states such as Japan and Germany could help the FSRs build MOX fabrication plants or modify nuclear reactors to burn plutonium in the FSRs, but Japan and Germany (and other nonnuclear weapon states) should not conduct these activities in their own countries. Otherwise, other nonnuclear weapon states, such as North Korea and Iran, might be unwilling to forgo sensitive civilian activities or plans.

The issue of burning FSR, as well as U.S., weapon-grade plutonium in the United States should be handled carefully, because the U.S. actions and intent can easily be misinterpreted. The United States should not seem motivated by self-interest or ask other countries to do the dirty work (i.e., to burn FSR plutonium). Otherwise, we have no objection to burning weapon-grade plutonium in the United States. On the other hand, we are reluctant to let nonnuclear weapon states such as Japan and Germany burn weapon-grade plutonium in their countries, because that would help these states further develop their plutonium use infrastructure, thereby defeating our general proposal of stopping or slowing plutonium use at least in nonnuclear weapon states.

An option we do not recommend is to place FSR weapon-grade plutonium under bilateral or IAEA safeguards in FSRs. The option is undesirable because no safeguards can prevent FSRs from reusing the materials for weapons in the event of drastic political change. If this option is adopted anyway, the parallel efforts of helping FSRs dismantle much of their nuclear weapon infrastructure assume added importance. Dismantling should make it as difficult, costly, and time-consuming as possible for FSRs to rearm themselves with a large number of nuclear weapons.

Chapter Five

DEALING WITH HIGHLY ENRICHED URANIUM FROM DISMANTLED NUCLEAR WEAPONS

This chapter begins with a discussion of the technical feasibility of blending HEU with depleted or natural uranium. The resulting low-enriched uranium (LEU) can be used to fuel current nuclear reactors. Then, we discuss how to calculate the value of HEU. We also estimate the potential market effect of LEU, blended down from HEU. Finally, we recommend a U.S. policy toward FSR HEU.

BLENDED-DOWN HEU

HEU in FSR and U.S. stockpiles exists in several forms, including metal for weapon production and as uranium hexafluoride (UF_6). Each form requires special handling.[1] Some naval, research, and high-temperature gas-cooled reactors use HEU. Such requirements, however, amount to a small fraction of the total HEU inventory. The bulk of HEU can be used as a feed to produce LEU.

Although the processing of natural uranium to LEU is a well-established technology, the United States and FSRs have essentially no experience de-enriching HEU to LEU. However, the process is quite feasible. A unique problem of de-enriching from HEU to LEU compared to enriching from natural to LEU is safety, or the possibility of creating a critical assembly. Special control measures must be em-

---

[1]Metallic uranium is extremely chemically active in the atmosphere, with most ceramic materials, and with other metals. As such, the processing of metallic uranium requires special handling.

ployed, such as batch size limitation, interaction spacing, the use of neutron-absorbing material, and special container geometry.

Although the de-enriching process is not performed now, the process is well understood. The licensing of de-enriching activities can follow the same procedures and regulations as those for licensing the HEU processing facilities. De-enrichment can be accomplished by blending the HEU with depleted or natural uranium, which is called the matrix material. The blending can be accomplished as a liquid UF_6 or a solid UO_2 powder.

For HEU in the form of UF_6, the preferred process is to blend it with the matrix UF_6 (hexafluoride with unenriched uranium). Two cylinders—a UF_6 cylinder of HEU and a UF_6 cylinder of matrix UF_6—would be heated to liquefy the UF_6. The UF_6 would then be metered from each cylinder (in the correct proportion to achieve the desired level of uranium enrichment) into a third, product cylinder. The product cylinder could then be either heated to receive the blend as a liquid or chilled to cause precipitation of the UF_6 as a solid.[2]

If the blend is in a liquid form, the filled product cylinder can be agitated (by rotation or shaking) to achieve concentration homogeneity in the blend.

After the production of the first batch of blend, its enrichment would be assessed and compared with that desired. Any needed adjustment of the enrichment would be accomplished by further blending.

For HEU in metal form, Mills[3] discusses a way to dissolve uranium metal in nitric acid, then purify the solution through solvent extraction, next convert it to UO_3, and finally reduce it to UO_2. The HEU in pure UO_2 powder is the product that would be supplied to the commercial fuel fabricators to blend it with the conventionally produced natural or depleted UO_2 powder. This blend would then be fabricated into fuel pellets.

---

[2]UF_6 blending techniques were discussed in standard texts at least as early as 1957. See, for example, Benedict and Pigford (1957), pp. 130–131, 148–152, 153–156, and 364–366. See also Mills (1992), and Benedict, Pigford, and Levi (1981), pp. 16–18, 225–229, and 919–921.

[3]Mills (1992) and Benedict, Pigford, and Levi (1981), p. 225–229.

The HEU in pure UO_2 powder that would be supplied to the commercial fuel fabricator presents a serious concern—the potential spread of weapon-usable HEU (as an oxide) into the marketplace.

An alternative to providing all of the HEU to the fabricator as low-enriched UF_6 can be achieved by steps of hydrofluorination of the UO_2 powder to form UF_4 and fluorination to form pure UF_6 for blending with matrix UF_6. A dry processing approach could also be used. In this approach, the solvent extraction purification step is eliminated. Consequently, an impure UF_6 will be formed. A fractional distillation step would be used to obtain pure UF_6.[4]

To address the potential safety problem related to forming a critical assembly, columns, piping, holding tanks, and blending tanks must be sized so that in the early stages of the de-enrichment process, less than 10 kg of HEU will be in residence at any one time and in any one location to eliminate the possibility of creating a critical mass. Working with smaller tanks and piping does present some engineering challenge because as the piping diameters get smaller, the likelihood of clogging increases. However, the engineering technology to overcome this potential problem is already in the facilities that have produced the HEU to begin with.

A HEU VALUE MODEL

Our model to determine the value of highly enriched uranium follows the same methodology as that for plutonium valuation. For a given amount of electricity generated, we determine the cost of using a current LWR without blended-down HEU. Then, we calculate the cost if blended-down HEU is used in the LWR. In the latter case, we include the cost of blending down the HEU to LEU. The cost difference with and without using HEU divided by the amount of HEU involved gives the value of HEU in $/kg.

In addition to data in Table 3.1, we have assumed that the blended-down cost consists of two components. The metal deconversion cost is $1,800/kg HEU and the blending cost is $1,000/kg HEU, for a total

---

[4]Mills (1992), p. 6.

of \$2,800/kg HEU.[5]   Further, we assume that FSR HEU has the same enrichment as that in the U.S. military, which is 93.5 percent enriched U-235.[6]

The value of surplus HEU by the year 2003 is shown in Table 5.1. FSR HEU is worth about \$6 billion and that in the United States \$3 billion at the current yellowcake price of \$10/lb U_3O_8. Thus, military HEU is valuable in civilian fuel cycles. Figure 5.1 shows the worth of FSR HEU at different uranium prices and discount rates.[7] The discounting applies to the 10-year period of nuclear weapon drawdown, as stipulated in START 2. We assume that the blended-down HEU is delivered and the money is received in 10 equal annual increments. Also shown in Figure 5.1 are the values of FSR weapon-grade plutonium as a function of uranium price.[8] Even at a yellowcake price of \$50/lb, more than five times the current price, the plutonium value is merely 8 percent that of HEU. It is a fortunate coincidence that the plutonium value is either negative or only a small fraction of the HEU value and that plutonium cannot be made safe but HEU can. This coincidence has significant implications on the U.S. policy toward FSR HEU and plutonium. When the United States deals with FSR HEU, the economic consideration should be of primary concern, because FSRs can earn substantially from HEU sales and HEU, once blended down, has little proliferation risk. When the United States deals with FSR plutonium, the United States should be less concerned about paying more than the plutonium's actual economic worth, since the overpayment would be fairly small. Instead, the United States should focus on the proliferation concern of plutonium and be generous.

---

[5]Steyn and Meade (1992), Appendix. Their plant is designed to handle 250 tonnes of HEU over its lifetime. Since we estimated that FSR surplus HEU by 2003 is 637 tonnes (see Table 2.1)—more than twice their figure of 250 tonnes—we are conservative in using their number unadjusted despite larger economies of scale.

[6]The U.S. military HEU is commonly known as oralloy and is enriched to 93.5 percent. Roser (1983), p. 4979.

[7]The value is determined using the assumption that uranium price is fixed throughout the 10-year period.

[8]We assumed that the plutonium will be used in LWR (FM, w/o R), where plutonium attains its highest value. See Figure 4.2.

Table 5.1

Total Value of Surplus HEU
(at $10/lb yellowcake and 10 percent discount rate)

	HEU (in tonnes)	HEU (in $ billions)
United States		
Warheads	210	2.0
Military inventory	129	1.3
Subtotal	339	3.3
FSRs		
Warheads	430	4.2
Military inventory	207	2.0
Subtotal	637	6.2
Total	976	9.5

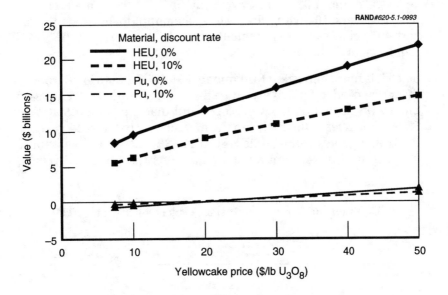

Figure 5.1—Comparison of Values of FSR HEU and Plutonium (637 tonnes of HEU versus 109 tonnes of plutonium)

POTENTIAL MARKET EFFECT OF BLENDED-DOWN HEU

By the year 2003, there will be a surplus of 339 tonnes of HEU in the United States and 637 tonnes in the FSRs (Table 2.1). If these materials were released to the commercial market, the utilities could use the blended-down HEU to fuel their reactors, thereby reducing their purchase of natural uranium and enrichment services worldwide.

To show the size of the effect of releasing these materials over the same 10-year period, 1993–2003, Table 5.2 shows the reduction in yellowcake and enrichment demand resulting from the release of a total of 976 tonnes of HEU. The reduction translates into 349,000 tons of yellowcake and 133 million of separative work units (SWU). Comparing these amounts with the current annual worldwide requirements of 71,000 tons of yellowcake and 29 million of SWU a year, we find that the release of HEU is equivalent to 4.9 years of yellowcake and 4.6 years of SWU supplies. If these materials are released over a 10-year period, they still amount to 49 percent of the current yellowcake requirement and 46 percent of the SWU requirement (Table 5.3).

Since it is more important to eliminate FSR HEU and to let them earn the much needed foreign exchange than for the United States to earn money on its own HEU, we recommend that the United States let FSRs alone release their HEU to the market over the next 10 years. If the United States withheld its HEU, the FSR HEU would amount to 32 percent of the current annual requirement in the world yellow-

Table 5.2

Uranium and Enrichment Displaced by Blended-Down HEU

	Yellowcake (tons)	Enrichment (millions of SWU)
United States		
Warheads	75,000	29
Military inventory	46,000	18
Subtotal	121,000	47
FSRs		
Warheads	154,000	58
Military inventory	74,000	28
Subtotal	228,000	86
Total	349,000	133

Table 5.3

Effect of Blended-Down HEU on World Uranium and Enrichment Markets

		Current Annual Uranium Requirement (tons of yellowcake)	Current Annual Enrichment Requirement (millions of SWU)
FSRs + Eastern Europe		12,000	5
Rest of the world		59,000	24
Total		71,000	29
Release over 10 years of	U.S.	17%	16%
blended-down HEU as a	FSRs	32%	30%
percentage of current			
annual requirement	Total	49%	46%

cake market and 30 percent in the enrichment market. Considering that utilities traditionally purchase a significant fraction of their requirement through long-term contracts, we believe that this is about the maximum amount that the markets can absorb.[9] To soften market disruption, we further recommend that the United States encourage itself and other countries, such as Japan and Germany, to purchase blended-down HEU and to keep natural or even enriched uranium for stockpiling. Many countries have long-standing concerns about the security of their energy supply. Stockpiling would help, but, perhaps more important, it would serve to take HEU off FSR hands, to help FSRs financially, and, at the same time, to reduce the disruption to the uranium and enrichment markets.

U.S. POLICY TOWARD FSR HEU

At the end of August 1992, the Bush administration announced that the United States had agreed in principle to purchase some 500

---

[9]Using 1991 data from the Western world, we estimated that only about 50 percent of uranium demand and as little as 20 percent of enrichment demand would remain unfilled over the next 10 years. Since a considerable amount of enrichment capacity in the world is still government owned, the government may have a better way to cut enrichment capacities temporarily. The source of 1991 data is Schreiber (1991), Figures 5 and 12.

tonnes of FSR HEU. The agreement calls for the purchase of 10 tonnes of HEU every year for the first five years and no less than 30 tonnes per year thereafter.[10] The Clinton administration has been pursuing the agreement further. In February 1993, William Burns, chief of the U.S. organization formed to help Russia dismantle nuclear weapons, said that the United States had agreed to purchase all of the HEU—roughly 500 tonnes—from FSR dismantled nuclear weapons over 20 years.[11]

The analysis here supports the policy of purchasing all FSR HEU or the LEU that derives from it. It further supports the idea of an orderly release of enriched uranium to the market. However, it is important to encourage FSRs to blend down their HEU in the military inventory as soon as possible. The same applies to HEU recovered from dismantled nuclear weapons. Although a country might want to save interest charges by delaying HEU blend-down until the LEU is needed, minimizing the risk of diverting HEU for nuclear bombs favors the start of blending operations as soon as practical. Moreover, the extra cost of early blending can be compensated by the savings of not having to store and safeguard the HEU as weapon-usable material.[12] We also recommend that the blending operation take place in FSRs instead of the United States or elsewhere. This would create some jobs for the FSR personnel involved in the nuclear weapon enterprise and other areas. Furthermore, domestic blending negates the need to transport HEU overseas.[13] Only if FSRs lack the capacity to blend HEU immediately, especially over the next few

---

[10]Reuters America Inc., September 1, 1992.

[11]As discussed earlier, FSRs may have considerably more HEU than 500 tonnes.

[12]Since the metal conversion and blending cost is about $2,800/kg HEU, the annual interest charge at 10 percent is $280/kg HEU. On the other hand, if we assume that the storage cost of HEU is one-third that of plutonium (because HEU's critical mass is three times as large), the storage cost of HEU without blending down would amount to about $140 to $290/kg HEU (or $330 to $670/kg HEU, if others' estimates are used) (see Table 4.1 and the discussion leading to Table 4.1). On this basis, the interest charges for immediate blend-down and for HEU storage seem comparable.

[13]Although transporting HEU is undesirable, we nonetheless recommend transporting FSR plutonium to the United States. The frequent transports of plutonium would indeed be worrisome, especially when users cannot afford heavy security for many trips. However, in our case, transporting FSR plutonium to the United States will involve only a small number of trips and the risk of long-distance transportation is more than compensated by the benefits of taking plutonium out of FSR hands.

years, should blending be done outside FSRs. The United States and other countries should provide financial and technical assistance to FSRs for building blending plants. The price of LEU from HEU can tie to the market prices (spot and long-term contracts) prevailing at the time of delivery or when the contract is signed.

Other countries should also participate in the purchase of blended-down HEU from FSRs. We are concerned that a U.S. commitment to purchase all FSR HEU may leave inadequate funds to also purchase their plutonium, especially given that FSRs are likely to have twice as much HEU as many have thought only recently.[14] Since we wish to prevent the spread of plutonium to nonnuclear weapon states and discourage any states from using it, it would be more advantageous for the United States to purchase FSR plutonium but encourage as many countries as possible to participate in the purchase of HEU, not vice versa.

---

[14]The cost to the United States of purchasing LEU from FSRs' 637 tonnes of HEU is $8 billion (discounted at 10 percent) to $12 billion (undiscounted). The HEU value to FSRs is $6 billion to $9 billion correspondingly, because they will have to spend $2 billion to $3 billion to blend their HEU into low-enriched uranium.

Chapter Six

FINDINGS AND RECOMMENDATIONS

There are inherent physical differences between uranium and plutonium. Once highly enriched uranium is blended down into low-enriched uranium, the process is not easily reversed and LEU cannot be used for nuclear weapons. But plutonium of any isotopic composition is weapon-usable and should be carefully controlled.

We found no economic benefits to fueling reactors with plutonium now. Our reference projection indicates that thermal recycle will not be economical until 50 years from now. Breeders will take even longer to become economical—100 years. Even under a very favorable projection, thermal recycle and breeders will still not be economical until 30 years and 50 years, respectively, from the present time. Moreover, since capital cost accounts for the major portion of the electricity cost, the use of plutonium in thermal reactors to save uranium and enrichment will not save much, even when thermal recycle becomes unexpectedly cheap. The use of breeder is further burdened by the higher capital cost. Most important, however, the widespread use of plutonium will greatly increase the risk of diversion by terrorist groups and seizure by host nations. There is little justification for introducing plutonium massively into the economy, when the benefits are nonexistent but the proliferation risk is real.

Plans for civilian nuclear power development in the 1950s and 1960s were concerned primarily with extending uranium resources and less with the proliferation implications of sensitive nuclear fuel cycles. By the 1970s, when the proliferation risk and the poor economics of thermal recycle and plutonium-based fast reactors became more recognized, many countries continued on the old course, because they worried that there was not enough time to develop more proliferation-resistant reactors and fuel cycles. The economics of plutonium use continued to worsen during the 1980s. More uranium ore was discovered in many places around the globe, and many nuclear projects were canceled worldwide. For example, all already-built nuclear power plants in the United States were ordered before 1975, or 18 years ago, and the picture is no brighter for the future. The increase in uranium supply and the decrease in uranium demand further postponed the day that plutonium would be needed. The 1990s began with the collapse of the communist empire, which brought forth a drastic reduction in nuclear weapons in FSRs and the United States.   The highly enriched uranium from dismantled nuclear weapons alone would provide an additional five years[1] of fuel supply for every reactor in the world. In addition, FSRs want to significantly increase their export of uranium and uranium enrichment.   These events have given countries enough time to develop proliferation-resistant reactors and fuel cycle arrangements. We propose, therefore, that countries reevaluate their civilian nuclear development policies and jointly develop a scheme for sharing the benefits of nuclear energy without increasing proliferation risk. The following agenda can help lead the world to that future.

- The United States should urge other countries to join in activities to terminate or drastically scale back plutonium separation and use.

- Countries should participate in programs that improve uranium efficiency in current once-through reactors; better assess uranium resources, especially those at higher prices; and develop advanced proliferation-resistant reactors.

- The United States should ask other countries to join in a plan to evaluate the feasibility and desirability of an international nuclear fuel cycle regime, in which the benefits, if any, of developing proliferation sensitive technologies will be shared among all countries; nonnuclear weapon states, in turn, would forgo the pursuit of these technologies. This evaluation will also be useful

---

[1]In fact, 8 years may be a better estimate, since the FSR HEU is likely to be about twice as much as the United States had recently thought.

in 1995, when countries consider how the NPT should be extended.

- If countries want to keep fissile materials for security reasons, or to assure a supply, the United States should encourage these countries to stockpile natural uranium, low-enriched uranium, and plutonium-containing spent fuel, instead of separated plutonium.

- As for military uranium, the United States should urge other countries, especially Japan and Germany, to share in the purchase of low-enriched uranium blended down from FSR highly enriched uranium.

- As for military plutonium, the first choice is to take FSR weapon-grade plutonium out of FSRs. A good way to accomplish that is for the United States, alone or with some help from the United Kingdom and France, to purchase all of it.

- If, after the best efforts of the United States and others, the FSRs still refuse to let weapon-grade plutonium leave their country, the second-best choice is to urge FSRs to burn it or dispose of it.

Fast Breeder Reactor   A fast breeder reactor contains no moderator and is simply an assembly of fissionable material and coolant contained within a structure. The fissionable material or fuel is cladded. The coolant includes sodium, a mixture of sodium and potassium, and helium. In fast reactors, the bulk of fission occurs at energies on the order of 100 keV. Fast breeder reactors breed more fuel than they consume.

Fast Mixed-Spectrum Reactor   The fast mixed-spectrum reactor uses both slow and fast neutrons to generate power and breed plutonium. It is designed to operate on a once-through-and-store fuel cycle, with fuel staying in the reactor for as long as 17 years and no fuel reprocessing. Only natural or depleted uranium is required for reload after the first core and transition cycles.

Fast Reactor   A fast reactor does not contain a moderator to slow down neutrons after they are generated. It is distinguished from a fast breeder reactor by not necessarily breeding more fuel than it consumes.

Fissile   The term "fissile" refers to nuclear materials that are fissionable by *both* slow (thermal) and fast neutrons. Fissile materials include U-235, U-233, Pu-239, and Pu-241. Materials such as U-238 and Th-232, which can be converted into fissile materials, are called fertile materials. It should be noted that Th-232, U-238, and all plutonium isotopes are fissionable by fast neutrons but not by thermal (slow) neutrons. They are not called fissile materials but may be called fissionable materials.

Fission   Fission occurs when a neutron bombards the nucleus of an atom and causes it to split into fragments and release energy.

Fissionable Material   Material whose nuclei fission when bombarded by neutrons.

Gigawatt Electric   A gigawatt electric (GWe) is equal to one thousand megawatts or one billion watts of electric power.

Heavy Metal   Heavy metal refers to all the isotopes of thorium, uranium, neptunium, plutonium, americium, and curium.

High-Level Waste   For this study, high-level waste (HLW) is the concentrate of the aqueous streams from the reprocessing cycles, which are used to extract plutonium and uranium from dissolved spent fuel.

Highly Enriched Uranium   For this study, highly enriched uranium is uranium with at least 90 percent fissile isotopes.

Light Water Reactor   There are two types of light water reactors (LWRs). One is a pressurized water reactor (PWR) and the other is a boiling water reactor (BWR). Both are thermal reactors. All commercially operating reactors in the United States and most commercial reactors worldwide are LWRs.

Light Water Reactor (Full MOX Fuel)   An LWR with full MOX fuel is fueled with fuel rods each containing a mixture or blend of uranium oxide and plutonium oxide. Traditional programs of using plutonium in LWRs start with partial, not full, MOX fuel.

Light Water Reactor (Once Through)   For this study, a once-through LWR is one that does not reprocess spent fuel or use any plutonium in the reactor's initial core and subsequent loadings.

Light Water Reactor (Partial MOX Fuel)   A LWR with partial MOX fuel contains some fuel rods that are blended with uranium oxide and plutonium oxide and some that are just uranium oxide. The blended uranium and plutonium oxides typically account for one-third of the total number of fuel rods.

Light Water Reactor (Without Reprocessing)   The spent fuel from a LWR can be reprocessed to recover plutonium and uranium. One concept, however, is to load the LWR with uranium or MOX (partial or full) but then not reprocess the spent fuel from that LWR. The plutonium for the MOX can be from dismantled nuclear weapons.

Liquid Metal Fast Breeder Reactor   A liquid metal fast breeder reactor (LMFBR) is one type of fast reactor.   An LMFBR uses no moderator and uses either sodium or a mixture of sodium and potassium as a coolant.

Low-Enriched Uranium.   Naturally occurring uranium contains only about 0.7 percent U-235 and almost all of the rest is U-238. Light water reactors typically use uranium fuel enriched to about 3.0 percent U-235.   Fuel enriched to about 3 percent or 4 percent is considered low-enriched uranium (LEU).

Mixed Oxide   Mixed oxide (MOX) refers to a physical blend of uranium oxide and plutonium oxide.

Pressurized Water Reactor   A pressurized water reactor (PWR) is a type of LWR whose primary coolant loops are not permitted to boil. The primary loops are typically under about 2,000 psi of pressure.

Prototype Fast Reactor   The first few fast reactors designed to demonstrate commercial use are prototypes.

Separative Work Unit (SWU)   The SWU is a measure of the work required to enrich uranium, and the SWU value depends on the U-235 concentration in the uranium one starts with (feed), the enrichment of the product (product), and the U-235 in the depleted waste (tails). For example, starting with natural uranium of 0.711 percent U-235, it takes 4.3 SWUs to produce 1 kg of 3 percent U-235 product with 0.2 percent tails.   It takes 3.4 SWUs to produce 1 kg of 3 percent U-235 product with 0.3 percent tails.

Thermal Reactor   Thermal reactors differ from fast reactors in that they use a moderator (such as water or graphite) to slow down the neutrons.

Thermal Recycle   Thermal recycle refers to the reprocessing of spent fuel and recycling the separated plutonium (with or without the separate uranium) as fuel into a thermal reactor such as a pressurized water reactor or a boiling water reactor.

Yellowcake   Yellowcake is the common name for uranium ore concentrates containing about 85 percent U_3O_8.

BIBLIOGRAPHY

Albright, David, Frans Berkhout, and William Walker, *World Inventory of Plutonium and Highly Enriched Uranium, 1992*, Oxford University Press, 1993.

Armed Forces Journal International, March 1992.

Augenstein, Bruno, "Some Aspects of Interstellar Space Exploration—New ORION Systems, Early Precursor Missions," IAA-91-716, 42nd Congress of the International Astronautical Federation, October 5–11, 1991, Montreal, Canada.

Benedict, M., and T. Pigford, *Nuclear Chemical Engineering*, McGraw Hill, New York, 1957.

Benedict, M., T. Pigford, and H. Levi, *Nuclear Chemical Engineering*, McGraw Hill, New York, 1981.

Berkhout, Frans, and William Walker, *Thorp and the Economics of Reprocessing*, Science Policy Research Unit, University of Sussex, Brighton, East Sussex, United Kingdom, November 1990.

Berkhout, Frans, et al., "Plutonium: True Separation Anxiety," *Bulletin of Atomic Scientists*, November 1992.

Bloomster, C. H., P. L. Hendrickson, M. H. Killinger, and B. J. Jonas, *Options and Regulatory Issues Related to Disposition of Fissile Materials from Arms Reduction*, DE91-008750, Pacific Northwest Laboratory, Richland, Washington, December 1990.

Broad, William, "Russian Says Soviet Atomic Arsenal Was Larger Than West Estimated," *The New York Times*, September 26, 1993, p. 1.

Carter, Ashton B., and O. Cote, "Disposition of Fissile Materials," *Cooperative Denuclearization: From Pledges to Deeds*, Center for Science and International Affairs, edited by Graham Allison, Ashton Carter, et al., John F. Kennedy School of Government, Harvard University, January 1993.

Chow, Brian, *The Liquid Metal Fast Breeder Reactor: An Economic Analysis*, American Enterprise Institute for Public Policy Research, Washington D.C., December 1975.

Chow, Brian, *Economic Comparison of Breeders and Light Water Reactors*, prepared for the U.S. Arms Control and Disarmament Agency, Pan Heuristics, 2805 Woodstock Road, Los Angeles, California 90046, July 1979.

Chow, Brian, *Timing of Breeder Demonstrations: An Economic Analysis*, Pan Heuristics, March 16, 1981.

Chow, Brian, *Off-Budget Financing for Clinch River: Higher Costs and Risks*, Pan Heuristics, Congressional testimony, September 15, 1983.

Cochran, Thomas, *The Liquid Metal Fast Breeder Reactor, An Environmental and Economic Critique*, Resources for the Future, Inc., Washington, D.C., 1974.

Combs, Jeff, "Uranium Supply and Demand Trends," Science Applications International Corporation, presented at the Fuel Cycle 91 Conference, Phoenix, Arizona, March 24–27, 1991.

"Construction Begun for Plant in Japan," *Nuclear News*, June 1993.

Department of Defense (DoD), *Military Plutonium from Commercial Spent Nuclear Fuel*, Congressional Hearings, Subcommittee on Oversight and Investigations, Commitee on Interior and Insular Affairs, U.S. House of Representatives, October 1, 1981.

Department of Energy, Office of Uranium Enrichment Assessment, "U_3O_8 Procurement, FY 1943–1971."

Department of State, *The International Control of Atomic Energy: Growth of a Policy*, Publication 2702, U.S. Government Printing Office, Washington, D.C., 1946.

El-Wakil, M. M., *Nuclear Power Engineering*, McGraw-Hill, New York, 1962.

Feiveson, Harold, and Theodore Taylor, "Security Implications of Alternative Fission Futures," *Bulletin of Atomic Scientists*, December 1976.

Ford-Mitre, *Nuclear Energy Policy Group Nuclear Power Issues and Choices*, Ballinger, Cambridge, Massachusetts, 1977.

"French Reprocessing," *Nuclear News*, January 1993.

Gillette, Robert, "Reactor Fuel Used in Bomb," *Los Angeles Times*, September 14, 1977a.

Gillette, Robert, "Impure Plutonium Used in '62 A-Test," *New York Times*, September 16, 1977b.

Glasstone, Samuel, and Alexander Sesonske, *Nuclear Reactor Engineering*, Van Nostrand Reinhold Company, New York, 1967.

Heritage Foundation, "Clinch River Breeder Reactor: A Portfolio of Charts," Washington, D.C., n.d.

International Atomic Energy Agency (IAEA), *IAEA Safeguards Glossary*, IAEA/SG/INF/1 (Rev. 1), 1987.

International Atomic Energy Agency, *IAEA Bulletin*, Vol. 34, No. 1, 1992.

Klemenic, John, "Examples of Overall Economics in Future Cycle of Uranium Concentrate Production for Assumed Open Pit and Underground Mining Operations," Uranium Industry Seminar, Atomic Energy Commission Grand Junction Office, Grand Junction, Colorado, GJO-108(72), October 1972.

Klemenic, John, "An Estimate of the Economics of Uranium Concentrate Production from Low-Grade Sources," GJO-108(74), October 1974.

Kouts, H.J.C., and W. Y. Kato, *The Fast Mixed Spectrum Reactor, Interim Report, Initial Feasibility Study*, Brookhaven National Laboratory, Upton, New York, December 1978.

Lamarsh, J. R., "Nuclear Reactor Theory," Addison-Wesley, New York, 1966

Leventhal, Paul, and Steven Dolley, *A Japanese Strategic Uranium Reserve: A Safe and Economic Alternative to Plutonium*, Nuclear Control Institute, Washington, D.C., April 12, 1993.

Lorie, J., and C. Gody, "Economic Analysis of Uranium Prices," prepared for Westinghouse Electric Corporation, July 9, 1975.

MacLachian, Ann, "Nuclear Stays France's Cheapest Power Even with Backend Costs," *Nucleonics Week*, May 27, 1993.

Mann, Jim, and Leslie Helm, "Japan Shifts Its Stand on Ruling Out A-Bomb," *Los Angeles Times*, July 9, 1993.

Military Plutonium from Commercial Spent Nuclear Fuel, Congressional Subcommittee on Oversight and Investigations, Committee on Interior and Insular Affairs, U.S. House of Representatives, October 1, 1981.

Miller, Marvin M., "Are IAEA Safeguards on Plutonium Bulk-Handling Facilities Effective?" Nuclear Control Institute, August 1990.

Mills, Loring, *Options for Commercial Use of HEU*, Fuel Cycle '92 Conference, Charleston, South Carolina, March 24, 1992.

Nuclear Energy Agency (NEA), *Plutonium Fuel: An Assessment*, Paris, France, 1989.

Nuclear Energy Agency, *Nuclear Energy Data*, 1992.

Nuclear Energy Agency and International Atomic Energy Agency, *Uranium: 1991 Resources, Production and Demand*, Paris, France, 1992.

Nuclear Engineering Handbook, McGraw Hill, New York, 1958.

Nuclear Engineering International (NEI), *World Nuclear Industry Handbook 1993*, New York, 1993.

Nuclear News, July 1992.

Nuclear News, June 1993.

Nuclear Regulatory Commission, *Generic Environmental Impact Statement for Mixed Oxide Fuels*, NUREG-0002, Washington, D.C., 1976.

NUEXCO, *Monthly Report on the Uranium Market*, Menlo Park, California, April 1982.

Partington, C., "Radiation Exposure Experience in Sellafield Reprocessing Plant," *Work Management to Reduce Occupational Dose*, OECD, Paris, France, 1993.

Roser, Herman, Assistant Secretary of Energy for Defense Programs in Department of Defense, *Authorization for Appropriations for Fiscal Year 1983, Part 7: Strategic and Theater Nuclear Forces*, Hearings, Washington, D.C., March 17, 1983.

Schlesinger, Jacob, "Japan Supports Open Extension of Nuclear Treaty," *Wall Street Journal*, September 28, 1993.

Schreiber, Kurt, "Eastern Europe's Market Role," Fuel Cycle 91 Conference, sponsored by U.S. Council for Energy Awareness, Phoenix, Arizona, March 24–27, 1991.

Selden, Robert W., *Reactor Plutonium and Nuclear Explosives*, Lawrence Livermore Natinoal Laboratory, Livermore, California, n.d.

Steyn, Julian, and Thomas Meade, *Potential Impact of Arms Reduction on LWR Fuel Cycle*, Energy Resources International, Inc., Washington, D.C., March 30, 1992.

Strauser, Wilbur A., Chief, Weapons Branch, Division of Classification, Energy Research and Development Administration, letter to Richard Bowen, Division of International Security Affairs, August 4, 1977.

Taylor, Theodore, "Nuclear Safeguards," *Annual Review of Nuclear Science*, Vol. 25, 1975.

Trutnev, Yu. A., and A. K. Chernyshev, *Report on the Fourth International Workshop on Nuclear Warhead Elimination and Nonproliferation*, Washington, D.C., February 26–27, 1992.

Vermeulen, François, "Combustibles Atomiques Nippons," *La Libre Belgique*, February 2, 1993.

von Hippel, Frank, *Report on the Fourth International Workshop on Nuclear Warhead Elimination and Nonproliferation*, Washington, D.C., February 26–27, 1992.

Wohlstetter, Albert, *The Spread of Military and Civilian Nuclear Energy: Predictions, Premises and Policies*, Pan Heuristics, 1976.

Wohlstetter, Albert, *Town and Country Planning Act 1971*, windscale testimony, Pan Heuristics, September 5–6, 1977.

Wohlstetter, Albert, et al., *Moving Toward Life in a Nuclear Armed Crowd?* prepared for the U.S. Arms Control and Disarmament Agency, Pan Heuristics, December 1975.

Wohlstetter, Albert, et al., *Swords from Plowshares*, University of Chicago Press, Chicago, Illinois, 1977.

The World Almanac, Scripps Howard Company, New York, 1991.

World Inventory [see Albright et al. (1993)].

Wright, John (ed.), *The Universal Almanac 1991*, Universal Press Syndicate Company, Kansas City, New York, 1991.